Movement as Medicine - Harnessing Physical Activity for Holistic Health across the Lifespan

Edited by Rashid Menhas

Published in London, United Kingdom

Movement as Medicine - Harnessing Physical Activity for Holistic Health across the Lifespan
http://dx.doi.org/10.5772/intechopen.1005927
Edited by Rashid Menhas

Contributors
Abdullah Aldughaysh, Aryan Shukla, Chinwendu Idoko, Christiana Panda, Daberechukwu Nwokedi, Dimitra Koutsouki, Eugénie d'Alessandro, Fredrick Isaac, Hafifi Hisham, Haleema Sadia, Hifza Ahmed, Ijeoma Ogunezi, Irtasam Ahmad, Jeneviv John, José Raúl Hoyos-Flores, Katerina Asonitou, Mohd Fahmi Aluwi, Monoem Haddad, Mupakile Chrispin, Nasser Abouzeid, Nor Azlin Mohd Nordin, Noriko Higuchi, Norma Angélica Borbón-Castro, Ogechi Obinka, Ricardo López-García, Rosa María Cruz-Castruita, Silvia Carolina Medrano-Mena, Sophia Charitou, Ulvina Riaz, Yusaku Sugiura

First published in London, United Kingdom, 2025 by IntechOpen
IntechOpen is the global imprint of INTECHOPEN LIMITED, registered in England and Wales, registration number: 11086078, 167-169 Great Portland Street, London, W1W 5PF, United Kingdom

For EU product safety concerns: IN TECH d.o.o., Prolaz Marije Krucifikse Kozulić 3, 51000 Rijeka, Croatia, info@intechopen.com or visit our website at intechopen.com.

British Library Cataloguing-in-Publication Data
A catalogue record for this book is available from the British Library

Movement as Medicine - Harnessing Physical Activity for Holistic Health across the Lifespan
Edited by Rashid Menhas
p. cm.
Print ISBN 978-1-83634-467-4
Online ISBN 978-1-83634-466-7
eBook (PDF) ISBN 978-1-83634-468-1

Meet the editor

Dr. Rashid Menhas is an Associate Professor at the Department of Nursing, Shandong Xiehe University, China, and a research fellow at Shinawatra University, Thailand. He has 6 years of postdoctoral research experience (Soochow University, China, from August 2019 to August 2022 and Zhejiang University, China, from November 2022 to April 2025). He is a social healthcare expert and medical sociologist focused on the psychosocial health and well-being of the population. His research has been published in numerous well-reputed international journals, and he has published more than twenty papers. He explores public health, physical activity, healthy aging, social prescription in primary care, e-learning, and behavioral health psychology. He is also an editorial board member of the journal "BMC Public Health."

Contents

Preface

In an era where sedentary lifestyles, mental health challenges, and chronic diseases loom large over global health, the transformative power of physical activity emerges as a beacon of hope. Movement as Medicine - Harnessing Physical Activity for Holistic Health across the Lifespan is born from the urgent need to reimagine health strategies that are inclusive, evidence-based, and grounded in the profound connection between the body and mind. This book is not merely a compilation of research but a call to action, a testament to the untapped potential of movement as a universal remedy for fostering resilience, equity, and vitality in diverse populations.

The human body is designed to move, yet modern life often conspires to keep us still. From screen-bound children to aging adults grappling with chronic conditions, the consequences of inactivity are stark: diminished mental health, fragmented social connections, and escalating healthcare burdens. Amidst these challenges, this book posits physical activity as a cornerstone of holistic well-being, a tool for healing, prevention, and empowerment that transcends age, ability, and circumstance.

Our exploration begins with the foundational role of physical activity in psychosocial health, weaving together insights from neuroscience, public health, and behavioral science. Chapters such as *The Healing Power of Movement – An Introduction to Physical Activity and Psychosocial Health* underscore how exercise shapes emotional resilience, cognitive function, and social bonds. These themes expand into discussions on special populations, where tailored interventions—like aqua therapy for mental resilience or multicomponent exercises for aging women—demonstrate that movement adapts to meet unique needs, whether in children with neurodevelopmental disorders or adults navigating physical disabilities.

The final section bridges theory with practice, offering actionable strategies for clinicians, educators, and policymakers. Guidelines for prescribing physical activity in chronic disease management, alongside evidence linking exercise to improved sleep and mental health, position movement as a viable and scalable intervention in both clinical and community settings.

This book is designed for a broad audience, including researchers seeking interdisciplinary connections, healthcare providers developing patient-centered care plans, educators promoting active lifestyles, and policymakers addressing health disparities. It also speaks to individuals and caregivers inspired to reclaim agency over their well-being. Each chapter is a piece of a larger mosaic, illustrating how movement fosters not only physical health but also dignity, joy, and a sense of belonging.

As you turn these pages, we invite you to envision a world where physical activity is accessible to all—a world where a child with autism finds confidence through sports, where an older woman rediscovers strength in community exercise, and where a college student balances screen time with restorative movement. Let this book be a guide, a catalyst, and a reminder: movement is not just medicine; it is a fundamental human right.

Together, let us step into motion.

Rashid Menhas
School of Medicine,
Department of Nursing,
Shandong Xiehe University,
Jinan, China

Section 1

Foundations of Psycho-Social Health and Well-Being

Chapter 1

Perspective Chapter: Essential Benefits of Physical Activity and Sports to Human Psychosocial Health and Well-Being

Mupakile Chrispin

"Lack of activity destroys the good condition of every human being, while movement and methodical exercise save it and preserve it". Plato (427–347 BC)

Abstract

Physical activity and sports are universally acknowledged for enhancing physical health, yet their profound contribution to psychosocial well-being and overall human wellness warrant equal emphasis. Engaging in regular physical activity has been linked to reduce stress, anxiety, and depressive symptoms, while elevating mood and self-esteem through neurochemical mechanisms such as endorphin release. Structured sports participation further fosters social cohesion, teamwork, and communication skills, nurturing a sense of belonging and strengthening interpersonal and community relationships. Cognitive benefits, including improved memory, concentration, and executive function, underscore the role of physical activity in enhancing academic and professional performance. Collectively, these psychosocial advantages promote emotional resilience, life satisfaction, mitigate cognitive decline and social isolation. Recognising the synergistic interplay between physical, psychological, and social health, this paper explores the essential benefits of physical activity and sports on physical, mental health, social well-being and personal development, emphasising their significance in fostering a balanced and fulfilling life. Policymakers, educators, and healthcare providers must prioritise accessible opportunities for physical engagement to address modern mental health challenges and cultivate holistic well-being in diverse populations.

Keywords: team, health, exercise, mental health, sports, diseases

1. Introduction

Physical activity is a crucial component of a healthy lifestyle, offering numerous benefits for physical and social well-being in all ages. It enhances cardiovascular health, strengthen muscles and bones, and reduces the risk of chronic diseases such as obesity, diabetes, and hypertension [1]. According to World Health Organization (WHO), physical activity plays an essential role in the prevention of non-communicable diseases.

Several research have indicated that regular physical activity impacts psychological resilience, mental well-being, and behavioural patterns. Less physical exercise and more inactive lifestyle are linked to health problems such as anxiety, weight gain, depression and other chronic conditions caused by sedentary behaviour [2].

According to WHO guidelines on physical activity and sedentary behaviour, 4 to 5 million deaths per year could be averted if the global population was more physically active [3]. Around one in two women and a third of all men in England are damaging their health through a lack of physical activity. An estimate of £7.4 billion is being spent every year in United Kingdom to manage the health care. If current trends continue, the burden of health and social care will destabilise public services and affect negatively the quality of life for individuals and communities [4].

During Covid-19 pandemic, the lives of many people around the world were affected. There was so much fear, and many were not sure if they were going to survive upon seeing many recorded cases of deaths especially in China, European and American countries. The scale of adversities caused a lot of stress, fear and depression thereby affecting the normal sleep patterns [5].

Additionally, the world population had spent more time on electronic communication devices and looking at screens, which prevented the practice of physical activities. This had the potential harmful long-term effects on chronic health conditions, including cardiovascular diseases, diabetes, cancer, and obesity all of which are known risk factors for severe COVID-19 [6].

The United States Physical Activity Guidelines recommend that all adults should engage in at least 150–300 minutes per week of moderate-intensity exercise whenever possible [7]. In addition, new evidence suggests that regular exercise may reduce the risk of acute respiratory distress syndrome, which is a leading cause of death in COVID-19-afflicted patients. This is mostly due to an increase in the production of extracellular superoxide dismutase during periods of exercise [6].

It is a proven fact that regular physical activities are key factors for the prevention and management of non-communicable diseases (NCDs) such as cardiovascular and respiratory diseases, various chronic conditions like obesity, type 2 diabetes, and cancer. Physical activity also benefits mental health, including prevention of cognitive decline and symptoms of depression and anxiety, and can contribute to the maintenance of healthy weight and general well-being [8].

Engaging in regular physical activities provide a wide range of benefits for the human body, affecting both physical, social and mental health. Physical activities play an important role in promoting people's health and well-being. Physically active individuals experience better sleep patterns, increased mental wellbeing and quality of life, as well as better physical function and fitness levels. In addition, physical activity might stimulate brain function and improve both cognition and mental health [9].

Regular physical activities like exercises have many physical health benefits, which are crucial in maintaining body functions. Physical activities such as martial arts training activities, running, cycling and swimming are known to improve heart and lung health by significantly reducing the heart disease by strengthening the heart muscle and increasing lung capacity [10].

Flexibility is also an important component that needs to be considered in maintaining physical health. This is a crucial component of physical fitness. Exercises such as yoga, Pilates and martial arts like karate are known to increase flexibility, which contributes to improved joint mobility, balance and posture. Regular stretching helps maintain the elasticity of muscles and tendons, allowing for smoother and more efficient movements [11].

Flexibility trainings can reduce injuries by loosening tight muscles and joints, especially in older individuals thereby preventing strains and sprains. Therefore, flexibility training is highly recommended as part of a daily exercise routine to maintain overall body health. Thus, sufficient regular exercise can maintain and improve physical health [12].

To ensure support towards physical activities, many countries like the United Kingdom ensured that every town has play parks to encourage people to do physical activities. Other individuals opt to subscribe to gyms with or without trainers to keep fit and healthy. According to the report by PlayTime Matters, children who are involved in physical activities tend to have increased self-confidence, social and environmental awareness as well as interest and curiosity in their surroundings. In addition, they also show great levels of creativity, resilience, problem solving [13].

In this modern era, the rapid development of technology and changing lifestyles have brought a huge impact on people's daily activity patterns. Many people now spend most of their time sitting in front of a computer screen or other electronic devices, leading to less physical activity [14]. This inactive lifestyle not only has a negative impact on physical health, but also on mental health.

Therefore, regular exercise is one of the important solutions to maintain a balance between mental and physical health. Regular exercise is proven to have many significant benefits for physical health. By exercising regularly, the body can improve fitness, strengthen muscles and bones, and improve flexibility and balance [14].

In addition, exercise also plays an important role in the prevention of diseases such as diabetes, hypertension and heart disease. These physical benefits gained from regular exercise can not only improve one's quality of life but also extend one's life expectancy [15].

2. Methodology

2.1 Research design

This article employs a qualitative research approach, utilising a literature review and observational analysis to explore the benefits of physical activities. The study synthesises existing research findings from peer-reviewed journals, health organisations, and expert opinions.

2.2 Data collection methods

The data for this article was gathered through:

Literature review: A comprehensive review of scientific studies, health reports, and books related to the benefits of physical activities. Sources include research articles from databases like PubMed, Google Scholar, and WHO reports.

Observational data: Analysis of trends and real-world cases showcasing the positive effects of physical activity on physical and mental well-being.

2.2.1 Data analysis

The collected data was categorised into key themes such as physical health benefits (e.g., Sports Participation and Physical activity, Benefits of physical activity in young children), mental health benefits (e.g., Role of physical activity in combating drug abuse, importance of physical activity in pregnancy), and social benefits (e.g., Social and economic benefits of physical activity). Findings were critically analysed and compared to identify consistent patterns and gaps in research.

2.2.2 Limitations

This study primarily relies on secondary data and self-reported information, which may have biases. Additionally, while it covers a broad range of benefits, it does not include all possible individual experiences with physical activity.

2.3 Sports participation and physical activity

Sports have been an integral part of human culture for millennia, serving not only as entertainment but also a vital tool for physical, mental and social development. Sport's value as a social connector is one of its most powerful development attributes. Sport is an inherently social process bringing together different fans, players, teams, coaches and other parties involved [2].

Sport creates extensive horizontal webs of relationships at the community level, and vertical links to national governments, sport federations, and international organisations for funding and other forms of support. In an era marked by sedentary lifestyles and rising health concerns, sports offer a dynamic solution to enhance quality of life [2]. Engaging in sports is one of the most effective ways to improve physical health. Regular participation strengthens the cardiovascular system, builds muscle and body density and enhances flexibility and coordination.

Sports is one of the physical activities that are demanding. Many sports activities have their specific training and require dedication. Sports like football, marathon or Basketball requires regular training to ensure individuals remain fit and ready for the challenge ahead. By training, the body become healthy and active thereby preventing diseases like diabetes and other diseases caused by sedentary lifestyle [1]. Emerging research highlights the critical role of physical activity in mitigating risk of diseases such as arthritis that come about due to excess body weight. Excess body weight is a significant risk factor for Osteoarthritis (OA), particularly in weight bearing joints like the knees and hips. Physical activities will assist in maintaining the healthy weight thereby reducing mechanical stress on joints [16]. Remaining healthy gives an assurance of a happy life without stressing of hospital bills or health insurance covers.

Sports like football and basketball elevate heart rate, improving circulation and reducing the risk of heart disease and stroke. According to WHO, physical inactivity is a leading factor for global mortality and sports can counter this by promoting heart health [3]. The physically inactive individuals with cardiovascular risk factors can reduce the risk of premature mortality by being active. The recommended 150 minutes of moderate aerobic activity has proven to prevent diseases [17].

2.4 Benefits of physical activities to young children

Physical activity is essential for children's overall health and development. Encouraging children to be active from a young age helps establish healthy habits that can last a lifetime. When an individual begins to exercise at an early age, exercising becomes part of their life. Without physical activity in children, their lifespan will be reduced as their chances of developing non-communicable diseases will be high [18]. For this reason, schools should always have provisions for outdoor activities that allows pupils to interact and perform certain physical activities. Not only will this be a motivation for physical fitness for the young ones but also improve their social interaction with people around them.

Physical education has become one of the most common activities incorporated in the school programmes worldwide. Children learn certain physical skills and discover their natural talents and capabilities that shapes their future. This regular participation in physical activities throughout childhood provides immediate and lifelong health benefits to the body [13]. School physical education (PE) provides a well-arranged physical activity participation that allows school going children to remain healthy and improve their social and psychological development.

Structured school programmes also mitigate risks associated with social anxiety. Extracurricular activities, such as drama clubs, or sports teams, provide platforms for children to explore interests, builds leadership skills, and form peer alliances. These experiences foster resilience by encouraging children to navigate success and failure in supportive environments. Additionally, research shows that children in schools with strong SEL (social and emotional learning) programmes exhibit greater prosocial behaviour, emotional stability, and academic engagement, underscoring the interdependence of psychosocial and cognitive development [19].

Schools integrating CASEL's SEL framework report better emotional regulations and peer relationships. For example, a Chicago school saw a 10% rise in cooperation after SEL integration [20]. SEL programmes assists children to develop the skills, knowledge, and attitudes necessary to understand and manage emotions. It also assists in setting and achieving positive goals, feel and show empathy for others, establish and maintain positive relationships, and make responsible decisions [21].

Schools that implement project-based learning for instance science fairs, art projects, group presentations foster a sense of accomplishment as students share ideas while working together to solve the given problem. Students in schools with Social and Emotional Learning (SEL) programmes showed improved self-efficacy and motivation [20].

While the efficacy of school programmes depends on equitable access and culturally responsive design. Children from marginalised backgrounds often face barriers to participation, exacerbating psychosocial disparities. Schools must prioritise inclusive policies, trained educators, and community partnerships to ensure all children benefit. By integrating psychosocial support into daily routines, schools lay the foundation for lifelong mental health and social competence [22].

2.5 Mental health benefits

Physical activity has numerous mental health benefits. Any increase in physical activity levels can result in a range of physical and mental health benefits. Engaging in regular activity can significantly improve mental well-being by reducing stress, anxiety, and depression while boosting mood, self-esteem, and cognitive functions [23].

Physical activities like exercises increases the brain-derived neurotrophic factor (BDNF) which promotes synaptic plasticity and neurogenesis especially in the hippocampus, which is involved in mood regulation. Since depression has been linked to reducing hippocampus volume, exercises increase the volume of BDNF thereby assisting in brain growth and better mood [24].

Regular physical activities can improve mood and sleep quality. Poor sleep exacerbates mental health conditions like depression and post-traumatic stress disorder (PTSD). Studies have shown that physical activities can improve sleep quality, duration and prevent the risk of developing disorders such as insomnia, daytime sleepiness and sleep apnea [25].

Being involved in physical activity gives assurance of a proper regulated circadian rhythm making it easier to sleep. Improved quality of sleep enhances mood, concentration and energy levels that benefits positively to mental health [26]. Research findings showed that, individuals who regularly exercise tend to have more regular and deeper sleep patterns, which is crucial for mental and physical recovery. Good quality sleep is crucial in the recovery process of the body and mind, making exercise an important component in maintaining a balance between mental and physical health.

Furthermore, being part of the physical activities, especially in groups, can increase a sense of community and social support. These interactions are very important in reducing feelings of social isolation that are often associated with mental disorders. By engaging in physical activity, individuals can build strong social networks, which in turn will improve emotional well-being and a sense of being part of the community [26].

3. The role of physical activity in combating drug addiction

Drug addiction is a complex disorder requiring multifaceted treatment approaches. Drug addiction is a significant public health issue affecting millions worldwide. Developing new medications to treat individuals affected with drug abuse is not only expensive, but also time consuming unsustainable [27]. Physical activity has emerged as a promising preventive and treatment therapy addressing biological, psychological and social aspects of addiction.

Substance abuse is a widespread problem and individuals experiencing substance abuse face challenges in the communities where they live. In most cases, individuals with abuse are seen to be violent and with unstable behaviour. In the UK, about 58–78% of the population believes that individuals with alcohol and drug addiction are dangerous, unpredictable. Not only is the public with this view but also the health professionals [28].

While most individuals face stigma in their society. Individuals have the fundamental need to be accepted and approved by the social group they intend to be part of. Group activities create collective objectives. Group activities like yoga, team sports, rock climbing and hiking give therapy treatment to individuals struggling with drug abuse disorders. These activities require teamwork, fostering trust and communication [29].

Overcoming addiction requires a multifaceted approach that includes medical intervention, psychological support, and lifestyle changes. Successful long-term recovery needs to involve experiences that are enjoyable and rewarding within a wider environment of new activities, behaviours and support [30]. In combating drug abuse, one of the most effective yet often overlooked tools is physical activity. Regular exercise has been shown to reduce cravings, improve mental health, and provide a healthier alternative to substance use.

Engaging in physical activities such as running, swimming, or strength training helps regulate brain chemistry, particularly dopamine and endorphins, the same neurotransmitters affected by drugs [31]. Exercise can naturally elevate mood and reduce withdrawal symptoms, making it easier for individuals to resist cravings. Studies have shown that aerobic exercise can significantly reduce drug cravings by decreasing stress and anxiety, which are common triggers for relapse. Activities like yoga and meditation-based workouts can also help manage withdrawal symptoms by promoting relaxation and mindfulness [31].

Many individuals struggling with drug addiction also battles with depression, anxiety, and low self-esteem. Physical activity has been proven to be a replacement of drug use by offering natural recovery remedy that eliminates the need to use drugs to get self-satisfaction [32]. When individuals with addiction commit to be involved in regular physical activities like exercises, this increases serotonin levels (neurotransmitter), which help improve mood and reduce stress. Physical activity like exercises provide a sense of accomplishment and control helping individuals rebuild their confidence whenever they achieve their fitness goals.

In addition, physical activities like exercises encourage individuals to develop healthy habits, replacing the time previously spent on drug use with meaningful and productive activities. Group fitness programmes, such as martial arts or team sports, also provide a sense of purpose and social support, reducing the likelihood of relapse. This involvement will protect an individual's brain structure as well as the safety of the public community as most drug abusers tend to be in criminal activities [33].

3.1 Importance in sexual life

Physical activity offers numerous sexual health benefits for both men and women. Regular exercise improves cardiovascular function, hormone regulation, mental health, and overall physical fitness, all of which contribute to better sexual performance and satisfaction. Regular exercises enhance circulatory health, which is essential for sexual function. A good blood flow helps maintain erections in men and increases sensitivity and arousal in women. In addition, running and swimming can reduce the risk of erectile dysfunction in men [34].

Physical activities play a crucial role in strengthening the pelvic floor, which directly impacts sexual health for both men and women. Strengthening the pelvic floor muscles through exercises like Kegels and squats enhances erectile strength in men and orgasm intensity in women. The pelvic floor muscles support the bladder, uterus (in women), prostate (in men), and rectum, and their strength and flexibility contribute to better sexual function and satisfaction. A strong pelvic floor enhances arousal, orgasm intensity, and erectile function in men [35].

Physical activities like exercises help regulate hormones such as cortisol, insulin and sex hormones, which play a role in sexual health. E.g., high cortisol levels due to chronic stress can negatively impact libido, while exercise help lower cortisol levels. Physical activity in sexual life will eliminate the use of Viagra, medicine used to treat erectile disfunction. This medication if misused by individuals has potential to cause sudden drop in blood pressure, leading to fainting or heart-related complications [36].

Engaging in regular physical activity can significantly enhance sexual health by improving circulation, hormone balance, mental well-being, and confidence. It is a natural and effective approach to preventing and managing sexual dysfunction [36].

3.2 Economic benefits

According to World Health Organization (WHO), the economic burden of physical inactivity is large. Globally, almost 500 million new cases of preventable non-communicable diseases were expected to occur between 2020 and 2030 with an estimated treatment cost of US$ 300 billion or around US$27 billion annually if there was no change in the prevalence of physical inactivity [1]. Such huge amount of money will adversely affect the lower and upper-middle-income countries. Most African countries depend on foreign aid and once their sponsor countries are affected, it is very difficult to manage diseases on their own.

Investing in physical activity as a preventative health measure can reduce the financial burden on both individuals and healthcare systems. A study published in the Journal of the American Heart Association estimated that if 20% more people engaged in regular exercise, the United States healthcare system could save nearly $67.5 billion over 20 years [37].

Government initiatives promoting physical activity also generate long-term savings. For instance, the countries like Canada and Netherlands known to have high national physical activity report lower per capita healthcare expenditures compared to nations with high sedentary lifestyles [1].

Nowadays, many workplaces have encouraged their employees to participate in physical activities by offering a certain contribution towards the gym monthly fees for their employees. This initiative provides benefits for employees and for the workplace. In Belgium, the headquarters of the National Social Security Office (NSSO) constructed a sports facility to promote physical activities for its employees. This was because of the high levels of inactivity found among workers due to lack of time and motivation [8].

To improve physical activities in workplace, there should be promotion of employees walking to colleagues' desks instead of using internal telecommunications or emails when requiring something. Workplaces may construct their car parks further away from the offices to allow individuals to walk to the site. Furthermore, workplaces can also organise weekly sports activities for instance football, netball or play matches against other local teams or within different department in the workplace. Issuing of incentives such as awards, small vouchers to individuals and teams participating physical activities with media coverage encourages other companies and well as other different individuals [8].

Additionally, companies that invest in workplace wellness programmes, including physical activity, experience a positive return on investment (ROI). A meta-analysis by Baicker et al. found that every dollar spent on workplace wellness programmes, medical costs decrease by approximately $3.27, and absenteeism costs fall by $2.73 [38].

At individual level, the economic return on physical activity is very significant. Exercising will help you reduce or prevent spending money on diseases or conditions that arise because of having a sedentary lifestyle. Physical activities will enable the body to be more active making it possible for an individual to be more productive [2]. A more productive individual usually receives praises, promotions, bonus as well as pay raises which will assist in handing easily financial problems thereby eliminating financial stress.

Engaging in physical activity reduces the likelihood of taking sick days, ensuring stead income without interruptions. With some workplaces offering overtime, a more fit and active employee will have the energy to work more hours than the one who is not thereby making more money. These physical benefits gained from physical activities like exercises can not only improve one's quality of life but also extend one's life expectancy [17].

There is much evidence that involvement in regular physical activity reduces the risk of dying prematurely [39]. It enhances the quality of life. Being healthy means having reduced healthcare costs and a better ability to enjoy healthy retirement life without financial strain [40]. Moreover, avoiding early retirement due to health issues ensures a longer earning period thereby boosting the retirement saving. A healthy employee may manage to retire at 65 years while those with healthy complications may retire early on medical grounds.

Physical exercising like cycling or walking while going for work saves a lot of money for families. In England, 2024 bus fares went up to £3 per strip and walking to the intended place may save an individual's money especially on a long run. Walking or cycling does not only save money on individuals but also makes the environment safer and healthy by reducing on the pollutants which automobiles produce [15]. This in return offers a guarantee to have a safe environment with people living healthy lives without being affected by pollutants associated with cancerous diseases, which in many cases is incurable.

3.3 Social benefits

Physical activity is widely recognised for its physical and mental health benefits, but its impact on social well-being is equally significant. Participating in physical activities encourages social interactions and bonding. Whether playing in sports team, group running, martial arts or hiking with friends, these physical activities create opportunities for individuals to connect with others who share similar interests. Participating in these activities such as sport enhances social cohesion by fostering a sense of belonging to a community or group. Engaging in group exercise programmes helps reduce loneliness and social isolation, especially among older adults [41].

Physical activities have social benefits that can positively impact people's lives including community involvement as well as sense of connection with others. Involving in activities such as martial arts, football, and fitness classes, or jogging with others usually builds connection and interactions, builds confidence, collaboration and communication abilities [42]. In addition, such activities promote meaningful team building as well as mutual respect among individuals. Social exercise combats isolation especially for older individuals who might not have regular social opportunities like young ones.

Physical activity fosters emotional intelligence by encouraging individuals to manage their emotions, practice empathy and develop resilience. Team sports teach individuals how to handle wins and losses gracefully, regulate emotions under pressure and understand the feelings of teammates and opponents. These experiences contribute to improved interpersonal relationships and emotional well-being [43]. People who feel lonely may feel more activated when they go for a dancing class or even a football match. The connection and communication that is made between the team helps such individuals to feel more loved and cared for.

Physical activities tend to reduce individual challenges associated with depression and loneliness. Some individuals release their frustrations, depression and anxiety through physical activities such as exercises. Depression tends to take away an individual's self-confidence [44]. Engaging in physical activity with other people can reduce the feeling of loneliness and isolation. Positive interaction during exercise release endorphins, improving mood and reducing stress, anxiety and depression. Exercising with others also make physical activity more enjoyable. Group activities will encourage an individual work extra harder than they would do when they are alone. This in the ned nurtures emotional resilience, fosters social bonds and enriches overall life satisfaction.

Many forms of physical activity, particularly team sports require cooperation, communication, and mutual support. Playing on a team helps individuals develop important social skills, such as conflict resolution, anger management, leadership, and collaboration [2]. These skills are transferable to other areas of life including school, work, and personal relationships, making teamwork-oriented physical activities valuable beyond the playing field. Encouraging participation in group physical activities such as exercises, sports and recreational activities can lead to a healthier, more connected and healthy society.

3.4 Benefits to public health

Physical activity plays a crucial role in promoting public health by reducing the risk of chronic diseases, improving mental health, and enhancing overall quality of life. In Africa, where both communicable and non-communicable diseases (NCDs) pose major health challenges, increasing physical activity can greatly contribute to improving public health. Physical activity helps regulate blood pressure, improve insulin sensitivity, and maintain a healthy weight [1].

A study in Africa found that individuals who engage in regular physical activity had a 30% lower risk of developing type 2 diabetes compared to inactive individuals [45]. Additionally, in Nigeria, community-based exercise programmes have successfully reduced hypertension prevalence among older adults [46].

Physical activity boosts the immune system, reducing susceptibility to infections such as tuberculosis and respiratory diseases [47]. In Ethiopia, studies have shown that physically active individuals have lower incidence in tuberculosis compared to sedentary populations [48]. Regular exercise also improves lung function, which is essential in managing diseases like asthma and chronic respiratory infections.

In Malawi, people faced problems of hypertension and diabetes mellitus in the urban cities such as Lilongwe and Blantyre. Despite having clinics to manage these diseases, the management of diabetes and hypertension has often been substandard with these diseases being poorly controlled due to inadequate drug stocks and lack of life-long therapy. These clearly shows how important it is to promote physical activity as a preventive measure to reduce the development of diabetes, obesity, hypertension and other non-communicable diseases.

3.5 Benefits of physical activities in pregnancy

Pregnancy is a transformative period in a woman's life, accompanied by significant physical and hormonal changes. Engaging in regular physical activity during pregnancy offers numerous health benefits for both the mother and the child. While concerns about safety may arise, research confirms that moderate exercise is not only safe but also essential for maternal and fatal well-being. Physical activities in pregnancy helps maintain cardiovascular fitness, strengthens muscles, and improves endurance, making it easier for the woman to cope with the physical demands of pregnancy and labour [49].

Women experience several conditions during pregnancy. These includes back pain, mood swings, and gestational diabetes which affects their normal functioning of the body. Conditions like back pain may be reduced through light approved exercises which must also be consulted with the physiotherapist or medical doctor. Being involved in physical activities like exercises will not only lower the chances of severe back pain during pregnancy but also reduce the chances of developing gestation diabetes [50].

Physical activities improve insulin sensitivity thereby limiting foetal exposure to hyperglycaemia. This reduces foetal insulin secretion, a driver of excessive growth. Physical activities like exercises promote glucose utilisation by skeletal muscles, preventing hyperglycaemia [51]. Exercise modulates foetal programming, potentially reducing childhood obesity risk. This could be the reason why offspring of active mothers' exhibit lower adiposity and improved insulin sensitivity [49].

Exercise modulates adipokines (leptin and adiponectin) which regulates foetal energy metabolism. Balanced leptin levels correlate with healthy birth weights. Furthermore, physical activity in pregnancy boosts antioxidant enzymes (glutathione peroxidase) thereby protecting against placental oxidative damage [52].

Other complications may include newborn child having low blood glucose levels (neonatal hypoglycaemia), chances of Perinatal death and a possibly having an overweight or obese child with the chances of developing Type 2 diabetes in later life. Pregnant women should limit the amount of time spent being sedentary. Replacing sedentary time with physical activity provides health benefits [53].

4. Conclusion

Physical activity is a cornerstone of a healthy lifestyle, contributing significantly to physical, mental, and social well-being. It reduces the risk of chronic diseases such as cardiovascular ailments, diabetes, and obesity while strengthening muscles, bones, and overall bodily functions [1]. Beyond physical health, regular exercise enhances cognitive abilities, improves sleep patterns, and alleviates stress, anxiety, and depression by stimulating neurotransmitters like serotonin and endorphins [24]. The benefits of an active lifestyle extend across all age groups, from childhood development to elderly care, ensuring longevity and improved quality of life. Despite these well-documented advantages, global physical inactivity remains a pressing public health concern, exacerbated by modern sedentary lifestyles and limited access to recreational spaces [8].

Beyond personal health benefits, physical activity plays a critical role in fostering social cohesion and economic stability. Participation in sports and group exercises enhances teamwork, leadership skills, and community engagement, reducing social isolation and promoting inclusivity [13]. Moreover, the economic implications of physical inactivity are significant, with billions spent globally on managing preventable diseases linked to sedentary behaviour. The World Health Organization [1] estimates that nearly $300 billion will be spent on treating non-communicable diseases caused by inactivity between 2020 and 2030. Governments and workplaces that invest in promoting physical activity experience reduced healthcare costs, increased productivity, and enhanced employee well-being [38]. Schools incorporating structured physical education programmes contribute to children's emotional resilience and academic success, laying a foundation for lifelong health [19].

To overcome barriers to physical activity, a collective approach involving policy-makers, educational institutions, healthcare providers, and communities is essential. Strategies such as building accessible recreational facilities, integrating exercise into daily routines, and promoting workplace wellness programmes can significantly improve global health outcomes [8]. While developed nations have successfully implemented policies to encourage active lifestyles, developing countries often struggle with resource constraints. A collaborative international effort is needed to ensure equal access to opportunities for physical activity worldwide. Emphasising its role in disease prevention, mental health improvement, and social well-being can pave the way for a healthier, more connected society [40].

The World Health Organization (WHO) has developed several policy frameworks aimed at promoting physical activity as a preventive health measure. These frameworks include the Global Action Plan on Physical Activity (GAPPA) 2018–2030, which provides strategic directions for the governments and stakeholders to create environments that encourage physical activity [2]. Implementing policy frameworks that promote active lifestyles is therefore crucial for reducing healthcare costs, improving public health and enhancing economic productivity.

To guide global and national efforts in increasing physical activity levels, WHO developed four key policy areas which namely, active societies, active people, active environments and active systems. Each of these areas provides a comprehensive approach to fostering a physically active population. However, their implementation in developing countries presents unique challenges and opportunities. For instance, African nations struggle to implement the "Active Societies" policy area with their traditional lifestyle and cultural norms that may discourage particularly women and older adults to exercise [54].

These challenges may be addressed through strengthening public awareness campaigns, and media engagements [2]. Though there is still limited government funding for national physical activity campaigns, integrating physical activity promotion into existing public health campaigns (e.g., anti-obesity, kidney day and diabetes preventions programmes) can maximise impact.

To have "Active Environments" means that facilities that facilitate physical activity should be available. These includes safe parks, pedestrian-friendly cities, cycling lanes, and accessible recreational facilities [8]. These are the facilities most developing countries lack adequately. Urban planning often prioritises motorised transport over pedestrian and cycling pathways, discouraging active commuting [15]. Public parks and sports facilities are often scarce or poorly managed. Government can collaborate with private investors and international donors to fund the construction of safe, accessible recreational spaces addressing all safety concerns especially for women and children.

Furthermore, when the safe recreational facilities are provided, opportunities for physical activity in everyday settings may increase including in schools, workplaces, and community centres. These facilities may be used either by paying or for free especially for government managed parks. This will encourage participation in physical activity by all ages thereby having "Active People." To have active people in the society, exercise should be integrated into daily routines rather than relying solely on recreational activities [2].

School based physical education (PE) programmes are often underfunded in low-income nations and people in societies cannot afford to pay for school trips. Ensuring that schools prioritise PE as part of the curriculum can foster lifelong active habits by using the available resources [13]. This can be done through community engagement with local sports clubs, dance groups, and cultural physical activities and games that require less resources. In workplaces, wellness programmes are uncommon in developing countries due to resource constraints. However, simple interventions such as promoting walking meetings, stair usage, and office exercise breaks can enhance employee health and productivity [38].

The final policy which provides for "Active Systems" highlights the need for strong governance, research, and policy coordination to ensure sustainable action on physical activity. The challenges that many developing countries are comprehensive national policies on physical activities. Incorporating physical activity into existing health policies, such as non-communicable disease prevention strategies, can drive systematic change [40]. Limited resources hinder large-scale interventions such as data collection, research and strengthening national data monitoring systems. Developing countries can seek assistance from organizations such as WHO, World Bank and regional health bodies to implement active lifestyles initiatives.

Funding

No funding source is reported for this study.

Declaration

No conflict of interest is declared by the author.

I would like to acknowledge my own efforts and dedication in exploring the benefits of physical activity. This book is the result of extensive research and passion for promoting a healthier lifestyle. I am also deeply grateful to my wife for her unwavering support, encouragement, and inspiration throughout this journey. Her belief in my work has been invaluable.

Author details

Mupakile Chrispin
Independent Researcher, United Kingdom

*Address all correspondence to: mupakilechrispin@gmail.com

References

[1] WHO. Global Status Report on Physical Activity. Geneva: Department of Health Promotion, World Health Organization; 2022

[2] WHO. Global Action Plan on Physical Activity 2018-2030: More Active People for a Healthier World. Geneva: WHO Press, World Health Organization; 2018

[3] WHO. WHO Guidelines on Physical Activity and Sedentary Behaviour. At a Glance. Geneva: WHO Press, World Health Organization; 2020a

[4] PHE. Everybody Active, Every Day. An Evidence-Based Approach to Physical Activity. London, United Kingdom: Public Health; 2014. Available from: https://assets.publishing.service.org.uk/media/5a7b595e40f0b60aaa293e6a/Everybody_active_Every_Day_Evidence_-based_approach_CONSULTATION_VERSION.pdf

[5] Wright L, Steptoe A, Fancourt D. Are adversities and worries during the Covid-19 pandemic related to sleep quality? Longitudinal analyses of 46,000 UK adults. PLoS ONE. 2021;**16**:1-18

[6] Tovani-Palone et al. Importance of healthy habits during the COVID-19 pandemic: Perspective and suggestions. Electronic Journal of General Medicine. 2022;**19**:1-5

[7] U.S Department of Health and Human Services. Physical Activities for Americans. 2nd ed. USA: U. S Department of Health and Human Services; 2019

[8] WHO. Promoting Physical Activity in the Workplace. Current Status and Success Stories from the European Union Member States of the WHO European Region. Denmark: WHO Regional office for Europe; 2020b

[9] Maldanado AS. Physical activity, physical fitness and exercise interventions for preventing human health and preventing and treating chronic conditions across the lifespan. International Journal of Environmental Research and Public Health. 2021:6-10

[10] Sattelmair et al. Dose response between physical activity and risk of coronary heart disease. A meta-analysis. American Journal of Public Health. 2011;**124**:789-795

[11] Behm DG et al. Acute effects of static stretching on muscle strength and power: An attempt to clarify previous caveats. Sports Medicine. 2016;**41**:1-11

[12] Bowman K et al. Dynamic Aging: Simple Exercise for Whole-Body Mobility. California: Propriometrics Press; 2017

[13] Prisk C. Playtime Matters. A Report Comparing Playtime Worldwide and Highlighting the Evidence Case for Making Playtime a Key Part of the School Day. London: Semble Technology Limited; 2019

[14] Owen et al. Too Sitting: The Population Health Science of Sedentary Behaviour. Vol. 38. USA: The American College of Sports Medicine; 2010

[15] Pucher J, Buehler R. Walking and cycling for healthy cities. Built Environment. 2010;**36**(4):391-414

[16] CDC. Obesity and Arthritis. 2020. Available from: https://www.cdc.gov

[17] Wen et al. Minimum amount of physical activity for reduced mortality

and extended life expectancy: A prospective cohort study. The Lancet, International Journal of Health, Medicine, and Sports. 2011;**378**:1244-1253

[18] Bailey R. Physical activity as an investment in human capital. Journal of Sport Management. 2006;**20**(3):323-334. Available from: https://www.researchgate.net/publication/236339428_Physical_ActivityAn_Underestimated_Investment_in_Human_Capital [Accessed: 17 February 2025]

[19] Jones et al. The Evidence Base for How We Learn: Supporting Students' Social, Emotional, and Academic Development. Washington, DC, United States: Aspen Institute; 2017. Available from: https://www.aspeninstitute.org/publications/evidence-base-learn/

[20] Durlak et al. The impact of enhancing students's social and emotional learning: A meta-analysis of school-based universal interventions. Child Development. 2011;**82**(1):405-432

[21] Borowski T. CASEL's Framework for Systemic Social and Emotional Learning. Measuring SEL Using Data to Inspire Practice. Chicago: University of Illlinois; 2019

[22] Eccles, Roeser. Schools as developmental contexts during adolescence. Journal of Research on Adolescence. 2011;**21**(1):225-241. DOI: 10.1111/j.1532-7795.2010.00725.x

[23] Kalfin et al. Significant positive impact of regular exercise on mental and physical health. International Journal of Health, Medicine and Sports. 2024;**2**:89-93

[24] Kandola et al. Physical activity and depression: Towards understanding the antidepressant mechanisms of physical activity. Neuroscience & Biobehavioral Reviews. 2019;**107**:525-539. DOI: 10.1016/j.neubiorev.2019.09.040

[25] Alnawwar et al. The effects of physical activity on sleep quality and sleep disorder: A systematic review. Cereus. 2023;**15**:1-5. DOI: 10.7759/cureus.43595

[26] Chennaoui et al. Sleep and exercise: A reciprocal issue? Sleep Medicine Reviews Journal. 2015;**20**:59-72

[27] Montoya D, Volkow N. IUPHAR review: New strategies for medication to treat substance use disorders. Pharmacological Research. 2024;**200**:1-8

[28] Boekel et al. Stigma among health professionals towards patients with substance use disorders and its consequences for healthcare delivery: Systematic review. Drug and Alcohol Dependence. 2013;**131**:23-35. DOI: 10.1016/j.drugalcdep.2013.02.018

[29] Theodorakis Y, Hassandra M, Panagiotounis F. Enhancing substance use disorder recovery through integrated physical activity and Behavioural interventions: A comprehensive approach to treatment and prevention. Journal of Brain Sciences. 2024;**14**:534. DOI: 10.3390/brainsci14060534

[30] McKay JR. Making the hard work of recovery more attractive for those with substance use disorders. Addiction. 2017;**112**:751-757. DOI: 10.1111/add.13502

[31] NIDA. The Role of Exercise in Preventing and Treating Drug Abuse. North Bethesda, United States: National Institute on Drug Abuse (NIDA); 2021

[32] Smith MA, Lynch WJ. Exercise as a potential treatment for drug abuse:

Evidence from preclinical studies. Frontiers in Psychiatry. 2012;**2**:1-10

[33] Duncan JR. Current perspectives on the neurobiology of drug addiction: A focus on genetics and factors regulating gene expression. ISRN Neurology. 2012;**2012**:24. DOI: 10.5402/2012/972607

[34] Khera M et al. Effects of aerobic exercise on erectile function: Systematic review and meta-analysis of randomized controlled trials. The Journal of Sexual Medicine, Department of Biostatistics, Johnson City, United States. 2023;**20**:1369-1375

[35] Frawley HC et al. Effect of pelvic-floor muscle training on male sexual dysfunction. Physical Therapy. 2017;**97**(6):1092-1101

[36] Verma RK et al. Toxic effects of sexual drug overdose: Sildenafil (Viagra). ARC Journal of Forensic Science. 2019;**4**:26-31

[37] Garcia L et al. Health care savings associated with increased physical activity among adults in the United States. Journal of the American Heart Association. 2016;**5**(9):e003614

[38] Baicker K et al. Workplace wellness programs can generate savings. Journal of Health affairs. 2010;**29**(2):304-311

[39] Carlson S. Percentage of deaths associated with inadequate physical activity in the United States. Preventing Chronic Disease. 2018;**15**:170354. DOI: 10.5888/pcd18.170354

[40] Ding D et al. The economic burden of physical inactivity: A global analysis of major non-communicable diseases. Lancet. 2016;**388**(10051):1311-1324. DOI: 10.1016/S0140-6736(16)30383-X

[41] Gow et al. Social support and successful aging. Journal of Gerontology: Psychological Sciences. 2017;**72**(2):379-389

[42] Eime RM et al. A systematic review of the psychological and social benefits of participation in sport for adults: Informing development of a conceptual model of health through sport. International Journal of Behavioral Nutrition and Physical Activity. 2013;**10**(1):135

[43] Bruner et al. Group cohesion and positive youth development in team sport athletes. Sport, Exercise, and Performance Psychology. 2014;**2**(4):219-227. DOI: 10.1037/spy0000017

[44] Dsouza et al. Biological connection to the feeling of happiness. Journal of Clinical and Diagnostic Research. 2020;**14**(10):1-5. DOI: 10.7860/JCDR/2020/45423.14092

[45] Peltzer, Pengpid. Physical activity and diabetes risk in South Africa. BMC Public Health. 2018;**18**(1):765

[46] Ogunmola et al. Hypertension reduction through community exercise in Nigeria. Journal of Cardiovascular Disease in Africa. 2021;**12**(1):45-52

[47] Nieman DC, Wentz LM. The immune-enhancing effects of exercise. Journal of Sport and Health Science. 2019;**8**(3):201-217

[48] Ayele B et al. Physical activity and tuberculosis risk in Ethiopia. International Journal of Infectious Diseases. 2020;**92**:50-57

[49] ACOG. American College of Obstetricians and Gynaecologists (ACOG) Committee Opinion No. 804: Physical Activity and Exercise during Pregnancy and the Postpartum Period.

Washington, DC: American College of Obstetricians and Gynecologists; 2020

[50] UK Chief Medical Officers' Physical Activity Guidelines. Physical Activity for Pregnant Women. Assets. UK: DHSC; 2019

[51] Barakat et al. Exercise during pregnancy protects against hypertension and macrosomia: Randomized clinical trial. American Journal of Obstetrics and Gynecology. 2016;**214**(5):649, e178-e188

[52] Seneviratne SN et al. Exercise in pregnancy: Effect on fitness, oxidative stress, and placental vascularization. Journal of Clinical Endocrinology and Metabolism. 2020;**105**(7):dgaa312

[53] Reyes-Molina et al. Association between the physical activity behavioural profile and sedentary time with subjective well-being and mental health in Chilean University students during Covid-19 pandemic. International Journal of Environmental Research and Public Health. 2022;**19**:2107. DOI: 10.3390/IJERPH19042107

[54] Bailey et al. Physical activity: An underestimated Investment in Human Capital? Journal of Physical Activity and Health. 2013;**10**:289-308

Chapter 2

Perspective Chapter: The Healing Power of Movement – An Introduction to Physical Activity and Psychosocial Health

Jeneviv John, Fredrick Isaac, Christiana Panda, Daberechukwu Nwokedi, Ijeoma Ogunezi, Ogechi Obinka and Chinwendu Idoko

Abstract

This chapter examines the multifaceted impact of physical activity on health, emphasizing its role in psychosocial well-being. Psychosocial health is the dynamic interplay between mental, emotional, social, and environmental factors, underscoring the inseparable link between physical and mental wellness. Various forms of physical activity and psychosocial health are explored alongside recommended physical activity guidelines and key physiological mechanisms through which exercise improves mood, reduces anxiety, and enhances mental health. The chapter delves into movement as a therapeutic intervention, exploring the relationship between movement and mental health, forms of movement therapy, and barriers to physical activity. It also highlights the social benefits of group activities, such as fostering community, support networks, and belonging. Real-life case studies illustrate how movement therapy helps individuals overcome trauma, build confidence, and improve their quality of life. The chapter concludes with practical strategies for integrating physical activity into daily routines for holistic well-being.

Keywords: movement therapy, physical activity, psychosocial health, preventive medicine, mental well-being

1. Introduction

1.1 Overview of movement, physical activity, and psychosocial health

Movement refers to any activity that occurs naturally, typically not undertaken to improve health or fitness, and does not demand significant discretionary time [1]. Movement is distinguished from physical activity in that it involves any bodily motion without necessarily aiming for a goal. Physical activity (PA), on the other hand, is defined as any bodily movement produced by the contraction of skeletal muscles that results in energy expenditure [2]. Examples of physical activity span occupational tasks, sports, exercise, household chores, and recreational activities [3]. Exercise represents a specific subset of PA that is planned, structured, repetitive, and intentionally aimed at improving or maintaining physical fitness.

While the physical benefits of PA are well established, such as reducing the risk of premature mortality and enhancing physical health, its psychological benefits are also increasingly recognized [4]. PA has been shown to positively impact mental health, including addressing serious mental illnesses such as schizophrenia, substance abuse, and eating disorders, with calls to integrate PA into mental health interventions [4, 5]. Psychosocial factors such as mental well-being, motivation, and social support also play a critical role in influencing participation in physical activities [5].

Specific types of physical activity such as group exercises and solo activities offer unique emotional and social benefits. Group activities, such as aerobics, dance classes, or team sports, foster social connections, enhance a sense of belonging, and provide mutual encouragement, which are pivotal in improving emotional resilience and reducing loneliness [6]. These interactions create an environment where individuals can share experiences, overcome challenges together, and receive feedback that promotes self-worth and reduces anxiety. Conversely, solo activities like running, swimming, or yoga offer opportunities for introspection and self-regulation. They enable individuals to disconnect from external stressors, focus on personal goals, and achieve mental clarity [6]. Both types of activities contribute to emotional health by reducing stress and fostering feelings of accomplishment.

The underlying psychological mechanisms further explain these benefits. Physical activity stimulates the release of endorphins and neurotransmitters such as serotonin and dopamine, which elevate mood and combat depression [7]. It also regulates the hypothalamic-pituitary-adrenal axis, thereby reducing physiological responses to stress. Moreover, activities that involve mindfulness or rhythm such as yoga or group dancing can enhance emotional regulation by fostering a meditative state and promoting cognitive focus [7]. In group settings, social interactions amplify these effects by providing emotional support and shared experiences that buffer against stress and improve overall well-being [8].

These dynamics underscore the importance of integrating physical activity into strategies aimed at enhancing psychosocial health. Also, tailoring interventions to include both group and solo activities can optimize emotional and social dimensions of well-being, thereby addressing diverse individual needs and preferences.

1.2 Historical perspectives

The connection between physical activity and health has been acknowledged since antiquity [9]. Ancient Greek philosophers like Plato and physicians like Hippocrates and Galen emphasized the importance of movement and exercise for maintaining good health. Plato stated, “Lack of activity destroys the good condition of every human being, while movement and methodical physical exercise can save it and preserve it.”

Eastern traditions also recognized the value of PA. Indian Vedic practitioners such as Susruta and Charaka, and Chinese physician Hua T’O incorporated physical exercise into their holistic health practices [10]. This Renaissance sparked renewed interest in human anatomy and exercise. Advancements continued into the eighteenth and nineteenth centuries with Johann GutsMuths introducing physical education in schools and Friedrich Jahn pioneering gymnastics in Germany. By the twentieth century, researchers like Johannes Lindhard and August Krogh advanced the scientific study of exercise physiology, including metabolic and cardiorespiratory responses to activity, cementing the field of exercise science.

1.3 Categories of physical activity

Physical activity is classified based on intensity, typically measured in metabolic equivalents (METs), heart rate, or breathing impact. This categorization aids in tailoring fitness plans and health recommendations.

1. Low-intensity activities: Involves minimal effort and is below 3 METs (metabolic equivalents). Examples: Gentle walking, stretching, household chores like washing dishes, or leisurely playing an instrument.

2. Moderate-intensity activities: Moderate activities range between 3 and 6 METs and elevate the heart rate slightly while still allowing conversation. Examples: Brisk walking, recreational swimming, dancing, and gardening.

3. Vigorous-intensity activities: These require six or more METs, significantly increasing heart rate and breathing, and making conversation difficult. Examples: Running, cycling above 10 mph, competitive sports, and high-energy aerobics.

1.4 Physical activity guidelines

The World Health Organization (WHO) Guidelines on PA and sedentary behavior provide evidence-based public health recommendations for children, adolescents, adults and older adults on the amount of physical activity (frequency, intensity, and duration) required to offer significant health benefits and mitigate health risks [11]. The guidelines provide evidence-based recommendations for different populations. It indicates that all adults should undertake 150–300 minutes of moderate-intensity, or 75–150 minutes of vigorous-intensity physical activity, or some equivalent combination of moderate-intensity and vigorous-intensity aerobic physical activity, per week. Among children and adolescents, an average of 60 minutes per day of moderate-to-vigorous-intensity aerobic physical activity across the week provides health benefits. The guidelines also recommend regular muscle-strengthening activity for all age groups. Adults should engage in muscle-strengthening activities at moderate or greater intensity that involve all major muscle groups on 2 or more days a week, as these provide additional health benefits. Additionally, reducing sedentary behavior is recommended across all age groups and abilities, although evidence was insufficient to quantify a sedentary behavior threshold.

According to the guidelines, engaging in any amount of PA is better than doing none, regardless of the population. For individuals not currently meeting recommended activity levels, even small amounts of physical activity can provide health benefits. It is advisable to start with low levels of activity and gradually increase the frequency, intensity, and duration over time. The health benefits of regular physical activity and reducing sedentary behavior outweigh potential risks, which can typically be managed by gradually increasing the amount and intensity of activity. Pre-exercise medical clearance is generally not required. Individuals who are not regularly active and have no contraindications can safely begin and progressively increase their activity levels without consulting a health care provider.

For those already engaged in moderate-intensity activity, transitioning to vigorous-intensity activity can be done incrementally without consulting a healthcare provider. However, individuals who experience new symptoms while increasing their activity levels should consult a health care provider.

1.5 Categories of psychosocial health

Psychosocial health comprises various interrelated dimensions essential for mental and social well-being. It is a new term to comprehend the already established factors involved in mental health and psychological well-being [12].

Key components include emotional health, which involves managing emotions and coping with challenges such as stress, resilience, and emotional expression. Social health focuses on building meaningful relationships and contributing to the community through support systems like friendships and active participation. Mental or cognitive health emphasizes intellectual growth, decision-making, and problem-solving, supported by learning and creative pursuits.

Additional dimensions include spiritual health, which is about finding purpose and meaning in life, often through beliefs, values, or practices like meditation or gratitude. Environmental health examines the interactions with physical surroundings, promoting safe and sustainable living practices. Occupational health ensures balance and satisfaction in professional roles, enhancing career fulfillment and managing work-related stress. Together, these dimensions form a holistic framework, underscoring the interdependence of mental, social, emotional, and environmental factors in achieving overall wellness.

1.6 Movement as a therapeutic intervention

1.6.1 Physiological benefits

Physical activity offers substantial health benefits across multiple domains. Repeated exposure to mechanical stress and gravitational forces through regular activity enhances physical strength, endurance, bone mineral density, and neuromusculoskeletal fitness, fostering a functional and independent lifestyle [13]. Additionally, PA is associated with improved sleep quality and mitigation of psychiatric disorders [13]. The body produces neurochemicals like opioids and endocannabinoids during physical activity, which are linked to pleasure, reduced anxiety, improved sleep, and diminished pain sensitivity [14]. Moreover, regular PA positively regulates the hypothalamus-pituitary-adrenal axis, reducing cortisol secretion and balancing hormones such as leptin and ghrelin [13].

1.6.1.1 Impact of physical activity on neurochemicals

Exercise influences the synthesis and modulation of neurotransmitters, including monoamine catecholamines. These neurotransmitters, released in the brain, innervate critical regions like the hippocampus, amygdala, and prefrontal cortex, enhancing cognition and memory [15]. Dopaminergic neurons in the substantia nigra and ventral tegmental areas project along pathways involved in exercise-induced neurocognitive benefits [16]. Physical activity also impacts serotonin (5-HT) synthesis, increasing tryptophan transport across the blood-brain barrier and boosting serotonin levels [17]. Furthermore, exercise reduces inflammation and stress hormones, optimizes catecholamine function, lowers cortisol levels, and increases brain-derived neurotrophic factor, which enhances neural health [15, 18].

1.6.1.2 Impact of physical activity on inflammation and stress hormones

Regular exercise has immunomodulatory effects such as optimizing catecholamine, lowering cortisol levels, which in turn reduces stress levels, and lowering systemic inflammation. Physical activity has been shown to increase plasma brain-derived neurotrophic factor [18].

1.6.2 Psychological benefits

Numerous psychological effects, such as self-esteem, cognitive function, mood, depression, and quality of life, have been studied [19]. According to general results, exercise enhances mood and self-esteem while decreasing stress tendencies, a factor known to aggravate mental and physical diseases [20]. Studies show that people who exercise regularly have a better frame of mind. However, it should be highlighted that a consistent link between mood enhancement and exercise in healthy individuals has not been established.

1.6.2.1 Reduction of anxiety and depression symptoms

Physical activity has shown potential in alleviating depressive symptoms, sometimes as effectively as traditional antidepressants [21]. This effect is linked to the release of endorphins—opioid-like compounds produced by the hypothalamus-pituitary system during intense physical exertion or emotional arousal. Endorphins are thought to reduce pain, elevate mood, and foster social bonding, thereby mitigating symptoms of depression and anxiety [22]. Known as the "endorphin hypothesis," this theory suggests that PA increases the production of these endogenous peptides, reducing feelings of hopelessness and worry while enhancing emotional well-being.

1.6.2.2 Improved self-esteem and cognitive function

Physical activity enhances body capability, improving body image by fostering strength, flexibility, coordination, and muscle tone, rather than emphasizing esthetics [23]. These physical improvements contribute to better self-esteem and stronger interpersonal relationships. Additionally, exercise induces structural brain changes, such as increased gray matter in the frontal and hippocampal region and reduced gray matter damage [24].

Exercise also boosts the release of neurotrophic factors enhances blood flow, improves cerebrovascular health, and optimizes glucose and lipid metabolism, supplying vital nutrients to the brain [25, 26]. These physiological changes collectively enhance cognitive function and overall mental well-being.

1.7 The mind-body connection: How movement influences mental health

1.7.1 The interconnection of mental and physical health

Mental health disorders like anxiety, depression, panic disorder, and posttraumatic stress disorder are significantly correlated with the existence of physical health diseases like diabetes, hypertension, arthritis, and heart disease [27, 28]. Individuals with mental illnesses, including schizophrenia, bipolar disorder, depression, and even anxiety disorders, have lower life expectancy and worse physical health, especially cardiovascular disease [28, 29].

The "mind-body connection" highlights the reciprocal relationship between mental and physical health, where each significantly influences the other [30]. Mental health disorders such as anxiety, depression, and schizophrenia are strongly associated with physical conditions like cardiovascular disease, diabetes, and hypertension [27, 28]. This relationship often exacerbates health disparities due to factors like genetics, lifestyle, treatment challenges, and social determinants such as poverty and exclusion [31]. Co-occurring mental and physical health issues can lead to difficulties in daily functioning, such as managing family responsibilities or employment, often resulting in further isolation and compounded health challenges [30].

The interplay between these domains can create a cyclical effect: untreated mental health issues may worsen physical conditions and vice versa. For example, depression can negatively affect metabolism, increasing the risk of type 2 diabetes, while the stress of managing diabetes can intensify depressive symptoms [32]. Chronic physical conditions causing persistent discomfort, such as fibromyalgia, are also linked to higher rates of depression, which in turn can hinder effective treatment adherence [33]. This bidirectional impact underscores the need for holistic health assessments and integrated treatment approaches.

Addressing the co-occurrence of mental and physical health issues requires better collaboration between general and mental health services. Screening is essential in both medical and psychiatric settings to ensure timely diagnosis and intervention, ultimately preventing the vicious cycle where one condition exacerbates the other. Recognizing this interconnection allows for more comprehensive care, improving outcomes for individuals with complex health needs.

1.7.2 Forms of movement therapy

Movement therapy integrates Eastern and Western mindful techniques to promote holistic healing of the mind, body, and spirit [34]. Defined as a therapeutic approach using psychomotor expression, it includes forms like dance therapy, which emphasizes emotional, cognitive, and physical integration [35]. Practices like rhythmic movements and dance date back millions of years, fostering community support, proactive health engagement, and sustainable wellness [35, 36]. Combining movement therapy with conventional medicine aligns with the World Health Organization's definition of health, addressing physical, mental, and social well-being. The benefits include improved strength, balance, cardiovascular health, mood, and cognitive function, along with reduced pain and stress, making it an effective tool for enhancing quality of life and preventing illness [34]. Common forms of movement therapies include:

1. Yoga: Originating in ancient India, yoga combines physical postures, breathing techniques, and sometimes meditation to enhance mental and physical health. Various yoga styles have emerged in the United States, emphasizing different aspects of the mind-body connection but rooted in ancient philosophy. These range from gentle Kundalini yoga to challenging handstand exercises. Yoga's extensive benefits for mental and physical health have made it a widely researched therapeutic intervention [37–40].

2. Pilates: Pilates is a low-impact workout based on holistic principles such as concentration, control, and core stabilization. It strengthens deep muscles through resistance-based exercises and comes in two main forms: reformer and mat Pilates. Research supports pilates for managing pain, improving function, and enhancing psychological health and fitness in diverse populations, including those with low back pain and pelvic floor dysfunction [41, 42].

3. Tai Chi: This Chinese martial art integrates breathing, meditation, and coordinated movements to promote balance, stress reduction, and overall well-being by restoring the body's Qi [43]. Its simplified modern forms make it accessible to all ages. Over 500 trials and 120 systematic reviews confirm tai chi's benefits for physical and mental health [44].

4. Qigong: Rooted in traditional Chinese medicine, qigong combines mindful breathing and fluid movements to circulate Qi [45]. It can be practiced in various forms—standing, sitting, or lying down—making it inclusive and accessible. Studies show qigong reduces depression, improves quality of life, and lowers fall risk. Integrative TCQ (tai chi and qigong) approaches have shown benefits for cancer patients, including reduced fatigue, pain, and inflammation [45].

5. Feldenkrais Method: Developed by an engineer and physicist, the Feldenkrais Method (FM) uses movement exploration to enhance somatic learning by distinguishing between comfortable and strained motions. Repetition improves movement patterns, promoting self-efficacy and long-term health benefits [46]. Though less researched than TCQ, FM has shown promise for pain management, motor behavior improvement, and maintaining independence in populations with disabilities [47, 48].

1.7.3 Integration of movement therapy in mental health

Movement therapy, or dance therapy, offers an alternative or complement to traditional talk therapy, which originated in Western and European cultures as a narrative-based approach to psychotherapy. In contrast, movement and dance have been used for centuries to express, manage, and process emotions [49].

Historically, dance played a vital role in community life, healing, and emotional expression. Prehistoric practices ranged from shamans performing ritual dances to African natives singing and dancing barefoot in communal circles, and Chinese peasants on the Loess Plateau playing the Ansai drums to relieve fatigue [50].

Dance movement therapy emerged in the early twentieth century as modern dance and psychotherapy evolved. Modern dance liberated dancers from rigid, repetitive movements, encouraging creative, emotional expression [51]. Psychologists, recognizing the deep connection between the mind and body, began exploring non-verbal therapies. This integration of dance and psychotherapy gained momentum during the socially open atmosphere of the twentieth century, which emphasized individuality, self-expression, and self-discovery. By the 1940s and 1950s, these developments laid the foundation for dance movement therapy as a formal therapeutic approach.

In 1966, the American Dance Therapy Association was established to professionalize the field. Its objectives included standardizing education and qualifications, publishing a professional journal, organizing international conferences, fostering global collaboration, and promoting research and widespread use of dance therapy.

1.7.4 Challenges and considerations

Movement therapists face barriers to recognition within mental health fields, including skepticism and lack of acceptance from non-body psychotherapists [49]. Professional growth remains hindered by misconceptions about dance therapy's legitimacy and its theoretical underpinnings. Addressing these challenges requires continued advocacy, education, and interdisciplinary collaboration.

1.8 Barriers to movement and physical activity

Various factors can hinder efforts to become physically active, and in this article, these barriers are categorized into physical and psychological aspects.

1.8.1 Physical barriers

1.8.1.1 Lack of access to facilities or safe environments for physical activity participation

The environment significantly impacts participation in physical activity [52]. Factors such as access to facilities, traffic congestion, poor urban planning, crime, unsafe neighborhoods, and pollution shape our willingness and ability to engage in physical activity.

Limited access to facilities and the high cost of sports activities often prevents individuals from rural areas from engaging in such activities [53]. In semi-urban areas, recreational spaces such as parks and sports arenas are scarce, and streets with heavy traffic further discourage outdoor activities [54].

1.8.1.2 Mobility limitations

Mobility is closely tied to independence, social connection, activity, and personal security [55]. Impaired mobility can significantly impact an individual's well-being, leading to isolation, fear of falling, prolonged bed rest, and dependence on others for basic tasks.

People with mobility limitations often experience discrimination in public spaces, including stares and unpleasant remarks [56]. Additionally, they may face a lack of accessible facilities, insufficient social support from family and community, and inadequate guidance on engaging in physical activity [56, 57].

1.8.1.3 Health conditions that hinders physical activity

The relationship between physical activity and health conditions is complex. While physical activity offers numerous benefits, certain health conditions can increase risks, particularly for inexperienced or sedentary individuals engaging in sudden, intense exercise [58].

Exercise restrictions may be classified as absolute or relative, depending on the health condition. Absolute restrictions apply to conditions such as unstable angina, systolic blood pressure exceeding 180 mmHg, diastolic pressure over 100 mmHg, uncontrolled arrhythmias, heart failure, unstable diabetes, deep vein thrombosis, and conditions accompanied by fever [59]. Relative restrictions, while less prohibitive, apply to cases such as severe hypertension, severe electrolyte imbalances, severe hyperthyroidism, aortic stenosis, and coronary artery stenosis [60].

1.8.2 Psychological barriers

Psychological barriers, such as depressive thoughts, weight stigma, heightened perceptions of disability, lack of motivation, and fear, are significant obstacles to engaging in physical activity [61]. Extensive research highlights the critical role these barriers play in limiting participation in physical activity [62–64].

1.8.2.1 Fear

Fear can be detrimental and traumatic, leading to mental and physical concerns [61]. Studies conducted have shown that fear-related problems can limit promoting healthy habits [61, 64, 65]. Fear of injury, judgment, pain, or failure can discourage individuals from participating in physical activity. Additionally, many feel threatened by new routines, public settings, and culture.

1.8.2.2 Motivation

Motivation governs and shapes the direction, intensity, and persistence of behavior, influenced by a wide range of internal, external, and social factors [66]. In sports psychology, two dominant theories explain motivational dynamics: the Self-Determination Theory and the Achievement Goal Theory [67].

The Achievement Goal Theory emphasizes the importance of goal-setting in shaping decision-making, engagement, and behaviors to achieve performance outcomes. For instance, an individual might engage in a specific physical activity purely to master or learn a new skill [68]. Conversely, the Self-Determination Theory differentiates between intrinsic motivation driven by interest, enjoyment, or competence and extrinsic motivation, such as goals related to fitness, health, attractiveness, or social connections [69].

Much of the research on motivation, participation, and choices in physical activity is grounded in the Self-Determination Theory [70–72].

1.8.2.3 Historical trauma or negative experiences with movement

Repeated success or failure in specific activities significantly influences an individual's self-worth in those tasks [73]. Over time, these experiences extend to related domains, shaping a broader sense of overall competence or incompetence.

Individuals who struggled, failed, or felt discomfort while engaging in physical activities during childhood, adolescence, or adulthood often internalize the belief that they are generally incapable of excelling in such activities. This perception reinforces their reluctance to participate in physical activity, creating a cycle of avoidance and diminished confidence [74, 75].

1.9 Overcoming the barriers to movement and physical activity

1.9.1 Strategies for improving accessibility

Creating environments that are both accessible and encouraging for physical activity across diverse populations is a global health priority [76, 77]. To support the adoption of healthy behaviors, individuals need access to programs and facilities that promote recreation and physical activity.

Community spaces such as gyms, sports facilities, and recreation centers play a crucial role by offering diverse opportunities for both organized and unstructured activities, whether conducted individually or in groups. Their proximity and variety of services make them particularly relevant for engaging populations with different needs and interests [78]. Participation in these facilities can be encouraged through partnerships, awareness campaigns, and financial incentives, such as discounted memberships. Additionally, technology, such as fitness applications and online classes, broadens access to physical activity. Addressing psychological barriers through motivational programs and beginner-friendly activities fosters a more equitable and welcoming environment for all participants.

1.9.2 Therapeutic approaches to address anxiety and build confidence

Exercise has been shown to help individuals manage anxiety by exposing them to anxiety-like sensations in a controlled, low-risk setting, thereby enhancing their tolerance for unpleasant emotions and high arousal [79].

High-intensity exercise, which produces transient emotions and interoceptive cues like an elevated heart rate, can help individuals develop positive associations with these sensations and reduce avoidance behaviors. Furthermore, exercise strengthens impulse control by prioritizing long-term goals over immediate discomfort and teaches techniques, such as controlled breathing, to moderate arousal. Gradual increases in exercise intensity enable participants to build resilience and improve their ability to manage challenging situations effectively.

1.10 Journeys toward healing with movement

Movement is a powerful tool for healing, offering pathways to improved physical, mental, and emotional well-being. For many, movement serves as a therapeutic outlet, one that helps individuals confront and overcome barriers, whether they stem from anxiety, trauma, or physical challenges. These personal stories demonstrate the transformative potential of movement in reshaping lives and building resilience. Each narrative highlights a unique journey toward recovery and growth through physical activity. From dance and yoga to running and hiking, these stories underscore the diverse ways in which movement can empower individuals to reclaim their bodies, process emotions, and find a sense of community and purpose.

1.10.1 Personal narrative 1: Overcoming anxiety through dance

1.10.1.1 A personal story of transforming through dance movement

Dance Movement Therapy, also known as dance therapy, was introduced by Marian Chace in the 1940s as a psychotherapeutic approach that uses movement and non-verbal expression to aid in emotional and psychological healing. The American Dance Therapy Association defines it as "the psychotherapeutic use of movement to promote emotional, social, cognitive, and physical integration of the individual, for the purpose of improving health and well-being" [80]; the European Association Dance Movement Therapy adds the "spiritual integration" to this list [81]. Dance therapy offers numerous benefits for both mental and physical health [82]. It has been shown to improve emotional well-being by reducing symptoms of depression, anxiety, and trauma. Additionally, it promotes social interaction, helps with stress reduction, and increases physical fitness. Research has shown that dance therapy has proven effective in addressing a range of medical conditions, including but not limited to anxiety, depression, schizophrenia, and Parkinson's disease [83–86].

Sveindís, who experienced symptoms of Generalized Anxiety Disorder from childhood and was officially diagnosed at the age of 20, shared how dancing transformed not only her physical health but also her mental and social well-being. She recounted:

"As I sit down in front of the computer in my little Finnish student apartment, Viola's words linger in my mind. Pole dancing has given me immense strength, flexibility I never thought I would ever have. It has shown me I am more capable than I dared to think I was. It has taught me I can do anything, as long as I practice. It has shown me that mistakes are only stepping stones toward success. It has given me an outlet, tools to better manage my anxiety, passion, and drive, friends in different parts of the world, and now it has given me my best friend. I've finally learned to embrace myself, I feel."

1.10.2 Personal narrative 2: Finding Community in Group Therapy

1.10.2.1 An account of building connections in a supportive exercise group

Exercising in a group context can significantly improve one's ability to sustain physical activity and enhance overall quality of life, compared to exercising alone [87, 88]. Group exercise classes have long been popular among gym-goers, even before the COVID-19 pandemic, and have adapted to virtual formats during and after the pandemic.

In a study employing diagonal narrative analysis, a collective story was shared by a group of disabled individuals who experienced exercise within a peer group. They reflected on the challenges of navigating gym environments that were often unwelcoming, yet how their collective determination and sense of community empowered them to stay active:

"We all want to be active but have experienced oppressive practices in the gym that have made this difficult. We understand the feeling of being invalidated in the gym, but we have fought to ensure we stay active and recognize it is not our failings but the failings of the gym environment. We will not let this stop us, and we will continue to exercise."

1.10.3 Personal narrative 3: Healing trauma via yoga practice

1.10.3.1 Individual experience of using yoga as a tool for reconciliation with past trauma

Therapeutic yoga involves the application of yoga postures and practices to treat various health conditions, aiming to prevent, reduce, or alleviate physical, emotional, and spiritual pain [89]. It incorporates yogic teachings to address structural, physiological, and emotional limitations. Gulden and Jennings [90] found that yoga's mind-body focus had numerous benefits, including promoting self-acceptance and spiritual growth, reducing trauma symptoms, and enhancing peace, empowerment, and self-compassion. Research has demonstrated that yoga improves muscular strength, flexibility, respiratory and cardiovascular function, aids recovery from addiction, reduces stress, anxiety, depression, and chronic pain, improves sleep patterns, and enhances overall well-being and quality of life [89].

Susan was sexually assaulted in her youth, and the sudden release of her attacker left her feeling unsafe and exposed. This traumatic experience triggered posttraumatic stress symptoms that severely affected her body, including poor digestion, weight gain, and hypervigilance. Reflecting on her experiences, Susan remarked, "Everything that's said about trauma and posttraumatic stress symptoms is shockingly true. From gaining weight to isolating myself, I stopped going out to work out, hiding from the world. It wasn't me—I'm not the kind of person to hide or become a recluse. I noticed how my body had changed, physically shrinking, my posture hunched in fear. Stretching revealed to me just how much I needed to reclaim my body and my life."

Susan's approach to yoga has evolved significantly since her focus shifted from purely physical fitness to integrating both the mental and physical aspects of healing. In the past, she practiced yoga primarily for the physical benefits, but now she views it as a tool for addressing the trauma she has experienced. This shift reflects how yoga can serve as a holistic practice that fosters emotional and physical restoration, helping individuals reconcile their body and mind [79, 90]. For Susan, yoga has become more than a workout. It is a path to healing, mindfulness, and self-empowerment.

1.10.4 Personal narrative 4: Recovery and resilience through running

1.10.4.1 A runner's perspective on using physical activity as a means of mental health recovery

Running serves as a powerful practice of mental and physical reflection, where the mind and body are united in motion. Through regular long-distance running, individuals often experience an increased awareness of both present and past thoughts, while the interaction between body and mind becomes clearer. Running, even for just 5–10 minutes a day at slow speeds, has been shown to significantly reduce the risks of death from all causes, including cardiovascular disease. Studies reveal that this small amount of running can lower the likelihood of mortality by up to 30%, with cardiovascular death risk decreasing by as much as 45% [91].

Ellie shared how running helped alleviate her depression, attributing the benefits to the release of endorphins. She explained, "I suffer a little bit from depression. This really helped make me feel, wow. This is a good feeling of endorphins." Reflecting on her experience with postpartum depression, Ellie noted how running gave her mental relief: "Running allowed me to burn off that anxiety but also to... relax and have time for myself, which has been huge." This time allowed her to clear her head, manage her anxiety, and find space for personal healing, highlighting the psychological benefits of physical activity in emotional recovery.

1.10.5 Personal narrative 5: Hiking as movement therapy

1.10.5.1 Exploring the healing power of nature through hiking

Studies are increasingly focused on the impact of outdoor recreation therapy on PTSD symptoms among veteras [92, 93]. Findings suggest numerous benefits, including enhanced attention, improved cognition and memory, better sleep quality, increased emotional stability, reduced stress and anxiety, heightened self-efficacy, and an overall improved quality of life.

One of the studies by Diertrich et al. [92] explored the effects of a six-month wilderness experience program for combat veterans with posttraumatic stress disorder. Participants highlighted the program's therapeutic value, with one respondent stating, "I accept myself more for who I am." One veteran noted the transformative impact, sharing, "I found myself not having to take my medications... I guess it's just natural medicine." These findings emphasize the profound role of nature-based therapies in supporting mental health recovery and fostering resilience.

1.10.6 *A quantitative study on the impact of physical activity on psychological well-being*

1.10.6.1 *Exploring the relationship between physical activity, motivation, and psychological well-being*

Granero-Jiménez et al. [94] carried out a study to investigate the relationship between physical activity, motivation, and psychological well-being in young adults.

A descriptive cross-sectional correlational study was conducted, utilizing a multiple linear regression analysis to assess these variables. An online survey measured participants' physical activity levels, motivation types, and psychological well-being. The final sample consisted of 489 young adults aged 18–35 years. Results revealed that higher levels of physical activity were positively correlated with better psychological well-being. Furthermore, motivation, particularly intrinsic motivation, emerged as a significant predictor of psychological well-being. This relationship was especially strong among male participants with higher activity levels.

This study highlights the substantial influence of physical activity on the psychosocial health of young adults and underscores the importance of fostering intrinsic motivation to enhance activity levels and improve mental health outcomes.

1.10.7 *A quantitative study on physical activity, hematologic indices, and quality of life of people living with human immunodeficiency virus*

1.10.7.1 *Exploring the relationship between physical activity, hematologic indices, and health-related quality of life*

John et al. [95] carried out a study to examine the relationship between PA levels, hematologic indices, and the domains of health-related quality of life among people living with HIV, highlighting the transformative potential of movement in the journey toward healing and resilience.

The study involved 80 participants recruited from the HIV clinic of a tertiary hospital. Participants' PA levels were assessed using the International Physical Activity Questionnaire, while health-related quality of life was measured using the RAND-36 questionnaire. Additionally, hematologic parameters, including hemoglobin, white blood cell count, and neutrophil levels, were obtained through laboratory analyses.

The findings revealed that as PA levels increased, participants exhibited a notable trend of improvement in hematologic parameters and health-related quality of life scores, reflecting the profound influence of physical activity on health. Specific domains of health-related quality of life, such as social functioning and pain management, were positively and significantly associated with PA levels.

This study underscores the healing potential of physical activity as a complementary therapy in the management of HIV. By enhancing HRQOL and fostering resilience, PA emerges as a critical tool in addressing the multifaceted needs of individuals living with HIV. Incorporating physical activity into treatment protocols can empower individuals living with HIV, fostering a holistic approach to their overall health and psychosocial well-being.

1.10.8 A quantitative study exploring the impact of physical activity on health and well-being in young adults with serious mental illness

1.10.8.1 Investigating the impact of physical activity on health and well-being in young adults with serious mental illness

This study by Hovland et al. [96] explores how young adults with serious mental illness perceive PA and its impact on their health and well-being. Through in-depth interviews with nine participants who engaged in an aerobic high-intensity interval training program, the research highlighted that PA was experienced as a meaningful and beneficial practice. The participants described PA as enhancing their sense of well-being and contributing to their overall health. They also noted that it played a significant role in increasing their mental strength and resilience. Thematic analysis revealed three key themes: PA led to positive changes in focus and well-being, it fostered mental strength, and the lack of social support and safety often prevented engagement in PA.

The findings underscore the importance of adapted PA in improving psychosocial health among young adults with serious mental illness. PA not only promoted a stronger sense of self-identity and mental well-being but also facilitated social engagement and stress management. However, the study emphasizes that for sustainable participation, it is essential for individuals to select activities that hold personal meaning and interest. Furthermore, social support and encouragement were identified as crucial factors in overcoming barriers to PA, suggesting that these elements are vital for fostering long-term engagement and mental health benefits for individuals with serious mental health.

1.11 The future of physical activity and psychosocial health

The connection between physical activity and psychosocial health continues to gain recognition as research underscores the positive impact of movement on mental well-being. With advancements in technology, PA is becoming more accessible, paving the way for innovative therapeutic approaches that build emotional resilience, alleviate stress, and enhance overall quality of life.

1.11.1 Emerging trends and advancement in technology and virtual support systems

It has been well-documented that PA is associated with better brain health and cognitive function in adults [93, 97, 98]. Despite evidence that supports a physically active lifestyle across the lifespan as important for healthy cognitive aging, the practical implementation of regular physical activity remains very challenging, which is shown by the number of adults who are unable to meet the recommended physical activity level [97, 99, 100]. Given the high prevalence of physical inactivity and the societal costs that cannot be ignored, the need to develop efficient intervention strategies to encourage physical activity became necessary.

Tele-exercise, which is the use of technology in physical activity intervention, is now in vogue [101]. This has become increasingly common in recent years due to new smart health and wellness devices and also the need to find flexible and suitable solutions to maintain an active and healthy lifestyle across age demographics [102]. Mobile apps that offer tailored workout programs on video, personalized programs, activity and nutrition tracking, live instructors, (group) video conferencing, wearable devices, sensors, and remote health monitoring are some of the range of options offered by tele-exercise [103, 104]. Technologies do not only offer flexibility as physical activity can be performed anywhere, whether indoor or outdoor, smart devices also collect data and show real-time information about the level of physical activity engaged in by users, as well as important parameters such as blood pressure, percentage of oxygen saturation, heart rate, temperature, and more [105]. Using this data, training programs are personalized as well as feedback, which helps to monitor their progress and effectively achieve fitness goals [106].

These technological trends incorporate the use of hardwares (wearables, smart workout equipment, etc.) and the use of software (smart analytics algorithms, AI-enabled applications, etc.) to provide a seamless experience. In this sense, virtual reality (VR) is an emerging tool that offers a quality experience, and through the interplay of its graphics processing and virtual training environments, there is a dynamic response to the user's experience. VR is also able to assist rehabilitation through integration with motion capture systems [107]. This equally incorporates gamification to ensure a higher physical activity level in some adults [108]. Zeng et al. [109] have found in their study that virtual reality-based exercise improves moods related to anxiety and depression.

1.11.2 The role of policy in promoting active lifestyles

1.11.2.1 Importance of public health initiatives and policies supporting physical activity

Physical activity and exercise have been recognized as vital to a person's life [110]. Its long-term advantages include the ability to improve overall physical and mental well-being. Additionally, physical activity also helps lower the risk of several diseases such as cancer, type 2 diabetes, cardiovascular diseases, etc. [111].

Given the clear implications, the World Health Organization (WHO) has been partnering with countries to provide evidence-based recommendations for PA. These documents usually contain information on the amount, type, and frequency of PA that will generally uplift the life of citizens and would cut across demographics such as children, adults, pregnant women, and the elderly [111].

Physical activity guidelines targeting individuals often fall short of delivering widespread benefits, as they fail to achieve sufficient participation levels across populations. This aligns with the current WHO recommendations, which emphasize a population-wide approach to physical activity, encouraging individuals to engage in at least 150–300 minutes of moderate-intensity aerobic activity or 75–150 minutes of vigorous-intensity activity weekly, alongside muscle-strengthening activities, to enhance both individual and public health outcomes. Therefore, public policies are better suited as they target a wider range of demographics and are effective as a response toward global physical inactivity level. The spectrum of influence is wider, and so is the impact [111]. These public policies are important because a 25% increase in PA level will help avert more than 1.3 million deaths yearly [112], and there is clear evidence that it is an effective strategy against numerous chronic diseases and reduces their risk by 20–30% [113].

1.11.3 Community-based approaches and their impact on mental health

Home-based and community-structured PA are becoming commonplace. Through the use of virtual reality and applications that provide real-time connectivity and feedback, there are structured programs for many people. The technologies can be used either to track daily PA levels, which are measured against those of peers in the neighborhood, or to be used in shared virtual reality games. The digital technologies allow for visualization of tasks (such as in videoconferencing or pre-recorded videos) and also to give and receive feedback.

Technologies have also been used to bring adults together, for instance, in green parks in the community, where physical activity are still performed using technology or without it. Snedden et al. [114] have shown that engaging in PA has positive psychological outcomes, including enhanced mental well-being and overall quality of life. Such community activities have also been shown to reduce anxiety and stress levels and the risk of depression [115, 116].

2. Conclusion

Physical activity is essential for both our physical and mental well-being. This chapter explored the profound interconnection between movement and mental well-being, underscoring its physiological, psychological, and therapeutic benefits. From promoting neurochemical balance and reducing stress hormones to alleviating symptoms of depression and anxiety, physical activity serves as a cornerstone of holistic health. Furthermore, movement therapies like yoga, tai chi, and dance therapy demonstrate the potential to bridge cultural and historical practices with modern mental health strategies. Practices such as yoga, tai chi, and dance therapy highlight how ancient traditions can be powerful tools for healing in modern life.

However, challenges like limited access to facilities, lack of motivation, and societal stigma can make it difficult for many people to stay active. Overcoming these barriers requires creative solutions, such as community programs, public health policies, and new technologies like virtual fitness platforms and apps that make exercise more accessible and engaging for everyone.

Looking ahead, it is clear that PA has a crucial role to play in building healthier, more resilient communities. By creating environments that support movement and integrating it into mental health care, people can live fuller, happier lives. Through continued advocacy, education, and innovation, the transformative power of movement can become a fundamental aspect of global health and wellness.

Acknowledgements

We would like to express our gratitude to the researchers, practitioners, and individuals whose work and personal experiences in movement therapy have informed and enriched this exploration of physical activity's role in promoting psychosocial health.

Conflict of interest

The authors declare no conflict of interest.

Author details

Jeneviv John[1]*, Fredrick Isaac[2], Christiana Panda[1], Daberechukwu Nwokedi[3], Ijeoma Ogunezi[1], Ogechi Obinka[1] and Chinwendu Idoko[1]

1 Department of Medical Rehabilitation, University of Nigeria, Enugu Campus, Enugu, Nigeria

2 Department of Physiotherapy, Federal Medical Centre, Umuahia, Abia, Nigeria

3 Department of Physiotherapy, Lagos University Teaching Hospital, Idi-Araba, Lagos, Nigeria

*Address all correspondence to: cynthia.john@unn.edu.ng

References

[1] Bennett H, Arnold J, Davison K. Exercising to improve movement quality: Why and how. ACSM's Health & Fitness Journal. 2021;**25**(3):20-27. DOI: 10.1249/FIT.0000000000000669

[2] World Health Organisation. Physical Activity. Geneva: World Health Organisation; 2016. Available from: http://www.who.int/topics/physical_activity/en/

[3] Caspersen CJ, Powell KE, Christenson GM. Physical activity, exercise, and physical fitness: Definitions and distinctions for health-related research. Public Health Reports. 1985;**100**(2):126-131

[4] Buckworth J, Dishman RK. Exercise Psychology. Champaign, IL: Human Kinetics; 2002

[5] Johannes C, Roman NV, Onagbiye SO, Titus S, Leach LL. Relationship between psychosocial factors and physical activity among undergraduate students from a south African university. International Journal of Environmental Research and Public Health. 2024;**21**(4):441. DOI: 10.3390/ijerph21040441

[6] Liu R, Menhas R, Saqib ZA. Does physical activity influence health behavior, mental health, and psychological resilience under the moderating role of quality of life? Frontiers in Psychology. 2024;**15**:1349880. DOI: 10.3389/fpsyg.2024.1349880

[7] Menhas R, Dai J, Ashraf MA, Noman S, Khurshid S, Mahmood S, et al. Physical inactivity, non-communicable diseases, and National Fitness Plan of China for physical activity. Risk Management Healthcare Policy. 2021;**14**:2319-2331. DOI: 10.2147/RMHP.S258660

[8] Saqib ZA, Dai J, Menhas R, Mahmood S, Karim M, Sang X, et al. Physical activity is a medicine for non-communicable diseases: A survey study regarding the perception of physical activity impact on health wellbeing. Risk Management Healthcare Policy. 2020;**13**:2949-2962. DOI: 10.2147/RMHP.S280339

[9] Sevilmis A, Özdemir I, García-Fernández J. The history and evolution of fitness. SPORT TK-EuroAmerican Journal of Sport Sciences. 2023;**12**(4):1-20. DOI: 10.6018/sportk.493851

[10] Tipton CM. The history of "exercise is medicine" in ancient civilizations. Advances in Physiology Education. 2014;**38**:109-117. DOI: 10.1152/advan.00136.2013

[11] World Health Organization. WHO 2020 Guidelines on Physical Activity and Sedentary Behaviour. Geneva: World Health Organization; 2020. Available from: https://apps.who.int/iris/bitstream/handle/10665/336656/9789240015128-eng.pdf

[12] Husain W. Components of psychosocial health. Health Education. 2022;**122**(4):387-401. DOI: 10.1108/HE-05-2021-0084

[13] Mahindru A, Patil P, Agrawal V. Role of physical activity on mental health and well-being: A review. Cureus. 2023;**15**(1):e33475. DOI: 10.7759/cureus.33475

[14] Godse AS, Shejwal BR, Godse AA. Effects of suryanamaskar on relaxation

among college students with high stress in Pune, India. International Journal of Yoga. 2015;**8**:15-21. DOI: 10.4103/0973-6131.146049

[15] Avery MC, Krichmar JL. Neuromodulatory systems and their interactions: A review of models, theories, and experiments. Frontiers in Neural Circuits. 2017;**11**:108. DOI: 10.3389/fncir.2017.00108

[16] Lin TW, Kuo YM. Exercise benefits brain function: The monoamine connection. Brain Sciences. 2013;**3**(1):39-53. DOI: 10.3390/brainsci3010039

[17] Patrick RP, Ames BN. Vitamin D and the omega-3 fatty acids control serotonin synthesis and action, part 2: Relevance for ADHD, bipolar disorder, schizophrenia, and impulsive behavior. The FASEB Journal. 2015;**29**(6):2207-2222. DOI: 10.1096/fj.14-268342

[18] Telles S, Singh N, Yadav A, Balkrishna A. Effect of yoga on different aspects of mental health. Indian Journal of Physiology and Pharmacology. 2012;**56**:245-254

[19] Biddle S. Physical activity and mental health: Evidence is growing. World Psychiatry. 2016;**15**:176-177. DOI: 10.1002/wps.20331

[20] Ghosh D, Datta TK. Functional improvement and social participation through sports activity for children with mental retardation: A field study from a developing nation. Prosthetics and Orthotics International. 2012;**36**:339-347. DOI: 10.1177/0309364612451206

[21] Babyak M, Blumenthal JA, Herman S. Exercise treatment for major depression: Maintenance of therapeutic benefit at 10 months. Psychosomatic Medicine. 2000;**62**:633-638. DOI: 10.1097/00006842-200009000-00006

[22] Dishman RK, O'Connor PJ. Lessons in exercise neurobiology: The case of endorphins. Mental Health and Physical Activity. 2009;**2**(1):4-9

[23] Alleva JM, Tylka TL. Body functionality: A review of the literature. Body Image. 2021;**36**:149-171. DOI: 10.1016/j.bodyim.2020.11.006

[24] Chaddock-Heyman L, Erickson KI, Holtrop JL, Voss MW, Pontifex MB, Raine LB. Aerobic fitness is associated with greater white matter integrity in children. Frontiers in Human Neuroscience. 2014;**8**:584. DOI: 10.3389/fnhum.2014.00584

[25] Hötting K, Schickert N, Kaiser J, Röder B, Schmidt-Kassow M. The effects of acute physical exercise on memory, peripheral bdnf, and cortisol in young adults. Neural Plasticity. 2016;**2016**:6860573. DOI: 10.1155/2016/6860573

[26] Mandolesi L, Gelfo F, Serra L, Montuori S, Polverino A, Curcio G. Environmental factors promoting neural plasticity: Insights from animal and human studies. Neural Plasticity. 2017;**2017**:7219461. DOI: 10.1155/2017/7219461

[27] Niles AN, Dour HJ, Stanton AL, Roy-Byrne PP, Stein MB, Sullivan G, et al. Anxiety and depressive symptoms and medical illness among adults with anxiety disorders. Journal of Psychosomatic Research. 2015;**78**(2):109-115

[28] Osby U, Brandt L, Correia N, Ekbom A, Sparen P. Excess mortality in bipolar and unipolar disorder in Sweden. Archives of General Psychiatry. 2001;**58**(9):844-850

[29] Fleischhacker WW, Cetkovich-Bakmas M, De Hert M, Hennekens CH, Lambert M, Leucht S. Comorbid somatic illnesses in patients with severe mental disorders: Clinical, policy, and research challenges. The Journal of Clinical Psychiatry. 2008;**69**(4):514-515

[30] Sewell CA, Martin JS, Abel WD. Psychiatric co-morbidity: Revisiting the "mind-body" connection. The West Indian Medical Journal. 2012;**61**(2):115-116

[31] O'Brien C, Gardner-Sood P, Corlett SK, Ismail K, Smith S, Atakan Z. Provision of health promotion programmes to people with serious mental illness: A mapping exercise of four South London boroughs. Journal of Psychiatric and Mental Health Nursing. 2014;**21**(2):121-127

[32] Tabák AG, Akbaraly TN, Batty D, Kivimäki M. Depression and type 2 diabetes: A causal association? The Lancet Diabetes and Endocrinology. 2014;**2**(3):236-245

[33] Singh G, Kaul S. Anxiety and depression are common in fibromyalgia patients and correlate with symptom severity score. Indian Journal of Rheumatology. 2018;**13**(3):168-172

[34] Phuphanich ME, Droessler J, Altman L, Eapen BC. Movement-based therapies in rehabilitation. Physical Medicine and Rehabilitation Clinics of North America. 2020;**31**(4):577-591. DOI: 10.1016/j.pmr.2020.07.002

[35] Mills LJ, Daniluk JC. Her body speaks: The experience of dance therapy for women survivors of child sexual abuse. Journal of Counseling and Development. 2002;**80**(1):77-85

[36] Hattori Y, Tomonaga M. Rhythmic swaying induced by sound in chimpanzees (pan troglodytes). Proceedings of the National Academy of Sciences. 2020;**117**(2):936-942

[37] Jeter PE, Slutsky J, Singh N, Khalsa SB. Yoga as a therapeutic intervention: A bibliometric analysis of published research studies from 1967 to 2013. Journal of Alternative and Complementary Medicine. 2015;**21**(10):586-592. DOI: 10.1089/acm.2015.0057

[38] Field T. Yoga research review. Complementary Therapies in Clinical Practice. 2016;**24**:145-161. DOI: 10.1016/j.ctcp.2016.06.005

[39] Mooventhan A, Nivethitha L. Evidence based effects of yoga in neurological disorders. Journal of Clinical Neuroscience. 2017;**43**:61-67. DOI: 10.1016/j.jocn.2017.05.012

[40] Wieland LS, Shrestha NS, Lassi ZS, Panda S, Chiaramonte D, Skoetz N. Yoga for treating urinary incontinence in women. Cochrane Database of Systematic Reviews. 2019;**2**(2):CD012668. DOI: 10.1002/14651858.CD012668.pub2

[41] Mazzarino M, Kerr D, Wajswelner H, Morris ME. Pilates method for women's health: Systematic review of randomized controlled trials. Archives of Physical Medicine and Rehabilitation. 2015;**96**(12):2231-2242. DOI: 10.1016/j.apmr.2015.04.005

[42] Fleming KM, Herring MP. The effects of pilates on mental health outcomes: A meta-analysis of controlled trials. Complementary Therapies in Medicine. 2018;**37**:80-95. DOI: 10.1016/j.ctim.2018.02.003

[43] Jahnke R, Larkey L, Rogers C, Etnier J, Lin F. A comprehensive review of health benefits of qigong and tai chi.

American Journal of Health Promotion. 2010;**24**(6):e1-e25. DOI: 10.4278/ajhp.081013-LIT-248

[44] Huston P, McFarlane B. Health benefits of tai chi: What is the evidence? Canadian Family Physician. 2016;**62**(11):881-890

[45] Wayne PM, Lee MS, Novakowski J, Osypiuk K, Ligibel J, Carlson LE, et al. Tai chi and Qigong for cancer-related symptoms and quality of life: A systematic review and meta-analysis. Journal of Cancer Survivorship. 2018;**12**(2):256-267. DOI: 10.1007/s11764-017-0665-5

[46] Hillier S, Worley A. The effectiveness of the feldenkrais method: A systematic review of the evidence. Evidence-based Complementary and Alternative Medicine. 2015;**2015**:752160. DOI: 10.1155/2015/752160

[47] Torres-Unda J, Polo V, Dunabeitia I, Bidaurrazaga-Letona I, García-Gil M, Rodriguez-Larrad A, et al. The Feldenkrais method improves functioning and body balance in people with intellectual disability in supported employment: A randomized clinical trial. Research in Developmental Disabilities. 2017;**70**:104-112. DOI: 10.1016/j.ridd.2017.08.012

[48] Brummer M, Walach H, Schmidt S. Feldenkrais 'functional integration' increases body contact surface in the supine position: A randomized-controlled experimental study. Frontiers in Psychology. 2018;**9**:2023. DOI: 10.3389/fpsyg.2018.02023

[49] Walsh MJ. Challenges Faced by Movement Therapists in Terms of Recognition of their Profession: A Project Based upon an Independent Investigation [Thesis]. Northampton, MA: Smith College; 2010. Available from: https://scholarworks.smith.edu/theses/515

[50] Duan X. Dance movement therapy - a massage for the soul. Dance Fashion. 2021;**3**:38-39

[51] Levy F. Dance and Other Expressive Art Therapies: When Words Are Not Enough. New York: Routledge; 1995

[52] Pedersen MRL, Bredahl TVG, Elmose-Østerlund K, Hansen AF. Motives and barriers related to physical activity within different types of built environments: Implications for health promotion. International Journal of Environmental Research and Public Health. 2022;**19**(15):9000. DOI: 10.3390/ijerph19159000

[53] Marcen C, Piedrafita E, Oliván R, Arbones I. Physical activity participation in rural areas: A case study. International Journal of Environmental Research and Public Health. 2022;**19**(3):1161. DOI: 10.3390/ijerph19031161

[54] Van den Berg L, Grobler WCJ. The influence of access to facilities on the physical activity level of high school pupils in Bophelong, a semi-urban area of South Africa. Mediterranean Journal of Social Sciences. 2014;**5**(23):905. Available from: https://www.richtmann.org/journal/index.php/mjss/article/view/4608

[55] Health. Vic. Mobility and Self-Care Support Independence, Victoria State Government. 2015. Available from: https://www2.health.vic.gov.au/hospitals-and-health-services/patient-care/older-people/falls-mobility/mobility/mobilty-independence [Accessed: November 30, 2024]

[56] Smith M, Calder-Dawe O, Carroll P, Kayes N, Kearns R, Judy Lin EY, et al. Mobility barriers and enablers and

their implications for the wellbeing of disabled children and young people in Aotearoa New Zealand: A cross-sectional qualitative study. Wellbeing, Space and Society. 2021;**2**:100028

[57] Ginis KAM, van der Ploeg HP, Foster C, Lai B, McBride CB, Ng K, et al. Participation of people living with disabilities in physical activity: A global perspective. The Lancet. 2021;**398**(10298):443-455

[58] Franklin BA, Eijsvogels TMH, Pandey A, Quindry J, Toth PP. Physical activity, cardiorespiratory fitness, and cardiovascular health: A clinical practice statement of the ASPC part I: Bioenergetics, contemporary physical activity recommendations, benefits, risks, extreme exercise regimens, potential maladaptations. American Journal of Preventive Cardiology. 2022;**12**:100424. DOI: 10.1016/j.ajpc.2022.100424

[59] Saiz LC, Gorricho J, Garjón J, Celaya MC, Muruzábal L, Malón MDM, et al. Blood pressure targets for the treatment of people with hypertension and cardiovascular disease. Cochrane Database of Systematic Reviews. 2017;**10**(10):CD010315. DOI: 10.1002/14651858.CD010315.pub2

[60] Basile C, Fucile I, Lembo M, Manzi MV, Ilardi F, Franzone A, et al. Arterial hypertension in aortic valve stenosis: A critical update. Journal of Clinical Medicine. 2021;**10**(23):5553. DOI: 10.3390/jcm10235553

[61] Hamer O, Larkin D, Relph N, Dey P. Fear-related barriers to physical activity among adults with overweight and obesity: A narrative synthesis scoping review. Obesity Reviews. 2021;**22**(11):e13307

[62] Myre M, Glenn NM, Berry TR. Exploring the impact of physical activity-related weight stigma among women with self-identified obesity. Qualitative Research in Sport, Exercise and Health. 2021;**13**(4):586-603. DOI: 10.1111/aphw.12187

[63] Cooper L, Ryan C, Ells LJ, Hamilton S, Atkinson G, Cooper K, et al. Weight-loss interventions for overweight/obese adults with chronic musculoskeletal pain: A mixed methods systematic review protocol. JBI Evidence Synthesis. 2016;**14**(5):57-67. DOI: 10.11124/jbisrir-2016-002725

[64] Hasan ABMN, Sharif AB, Jahan I. Perceived barriers to maintain physical activity and its association to mental health status of Bangladeshi adults: A quantile regression approach. Scientific Reports. 2023;**13**(1):8993. DOI: 10.1038/s41598-023-36299-7

[65] Tabrizi JS, Doshmangir L, Khoshmaram N. Key factors affecting health promoting behaviors among adolescents: A scoping review. BMC Health Services Research. 2024;**24**:58. DOI: 10.1186/s12913-023-10510-x

[66] Shizgal P. Motivation. In: Wilson R, Keil F, editors. The MIT Encyclopedia of the Cognitive Sciences. Cambridge, MA: MIT Press; 2001. pp. 566-568

[67] Roberts GC, Nerstad CG. Motivation: Achievement goal theory in sport and physical activity. In: The Routledge International Encyclopedia of Sport and Exercise Psychology. London, New York: Routledge; 2020. pp. 322-341

[68] Hsu RMCS. Motives for Practice of Physical Activities: An Evolutionary Approach [Thesis]. São Paulo, SP: University of São Paulo; 2022

[69] Ryan RM, Deci EL. Self-determination theory and the facilitation of intrinsic motivation, social

development, and well-being. American Psychologist. 2000;**55**(1):68-78. DOI: 10.1037//0003-066x.55.1.68

[70] Borges JC, de Oliveira Filho GG, de Lira CAB, da Silva RAD, Alves EDS, Benvenutti MJ, et al. Motivation levels and goals for the practice of physical exercise in five different modalities: A correspondence analysis. Frontiers in Psychology. 2021;**12**:793238. DOI: 10.3389/fpsyg.2021.793238

[71] Jenkins M, Houge Mackenzie S, Hodge K, Hargreaves EA, Calverley JR, Lee C. Physical activity and psychological well-being during the COVID-19 lockdown: Relationships with motivational quality and nature contexts. Frontiers in Sports and Active Living. 2021;**3**:637576. DOI: 10.3389/fspor.2021.637576

[72] Varella MAC. Evolved features of artistic motivation: Analyzing a Brazilian database spanning three decades. Frontiers in Psychology. 2021;**12**:769915. DOI: 10.3389/fpsyg.2021.769915

[73] Reynes E, Dumoulin C, Robert B, Berthouze SE. Why aren't they involved in physical activities? The hypothesis of negative self-perception due to past physical activity experiences. Cogent Psychology. 2019;**6**(1):1570691

[74] Allen B, Canuto K, Evans JR, Lewis E, Gwynn J, Radford K, et al. Facilitators and barriers to physical activity and sport participation experienced by aboriginal and Torres Strait islander adults: A mixed method review. International Journal of Environmental Research and Public Health. 2021;**18**(18):9893. DOI: 10.3390/ijerph18189893

[75] Rodriguez Espinosa P, Garcia LC, Vasquez JJ, Xiao L, Stafford RS, Day Krenzel L, et al. Understanding historical trauma among urban indigenous adults at risk for diabetes. American Indian and Alaska Native Mental Health Research. 2022;**29**(3):43-70. DOI: 10.5820/aian.2903.2022.43

[76] Kohl HW, Craig CL, Lambert EV, Inoue S, Alkandari JR, Leetongin G, et al. The pandemic of physical inactivity: Global action for public health. The Lancet. 2012;**380**(9838):294-305

[77] World Health Organization. Global Action Plan on Physical Activity 2018-2030: More Active People for a Healthier World. Geneva: World Health Organization; 2018

[78] Smith BJ, Mackenzie-Stewart R, Newton FJ, Manera KE, Haregu TN, Bauman A, et al. Twelve-month findings of the MOVE Frankston randomised controlled trial of interventions to increase recreation facility usage and physical activity among adults. PLoS One. 2021;**16**(7):1-15

[79] Smith PJ, Merwin RM. The role of exercise in management of mental health disorders: An integrative review. Annual Review of Medicine. 2021;**72**(1):45-62

[80] ADTA American Dance Therapy Association. 2018. Available from: www.adta.org

[81] EADMT European Association Dance Movement Therapy. 2018. Available from: www.eadmt.com

[82] Koch SC, Riege RFF, Tisborn K, Biondo J, Martin L, Beelmann A. Effects of dance movement therapy and dance on health-related psychological outcomes. A meta-analysis update. Frontiers in Psychology. 2019;**10**:1806. DOI: 10.3389/fpsyg.2019.01806

[83] Kaltsatou A, Mameletzi D, Douka S. Physical and psychological benefits of

a 24-week traditional dance program in breast cancer survivors. Journal of Bodywork and Movement Therapies. 2011;**15**:162-167. DOI: 10.1016/j.jbmt.2010.03.002

[84] Priebe S, Savill M, Wykes T, Bentall RP, Reininghaus U, Lauber C, et al. Effectiveness of group body psychotherapy for negative symptoms of schizophrenia: Multicentre randomised controlled trial. The British Journal of Psychiatry. 2016;**209**:54-61

[85] Rodrigues-Krause J, Farinha JB, Krause M, Reischak-Oliveira Á. Effects of dance interventions on cardiovascular risk with ageing: Systematic review and meta-analysis. Complementary Therapies in Medicine. 2016;**29**:16-28. DOI: 10.1016/j.ctim.2016.09.004

[86] Karkou V, Meekums B. Dance movement therapy for dementia. Cochrane Database of Systematic Reviews. 3 Feb 2017;**2**(2):CD011022. DOI: 10.1002/14651858.CD011022.pub2

[87] Kanamori S, Takamiya T, Inoue S, Kai Y, Kawachi I, Kondo K. Exercising alone versus with others and associations with subjective health status in older Japanese: The JAGES cohort study. Scientific Reports. 2016;**6**:39151. DOI: 10.1038/srep39151

[88] Stødle IV, Debesay J, Pajalic Z, et al. The experience of motivation and adherence to group-based exercise of Norwegians aged 80 and more: A qualitative study. Archives of Public Health. 2019;**77**:26. DOI: 10.1186/s13690-019-0354-0

[89] Woodyard C. Exploring the therapeutic effects of yoga and its ability to increase quality of life. International Journal of Yoga. 2011;**4**(2):49-54. DOI: 10.4103/0973-6131.85485

[90] Gulden AW, Jennings L. How yoga helps heal interpersonal trauma: Perspectives and themes from11 interpersonal trauma survivors. International Journal of Yoga Therapy. 2016;**26**(1):21-31. DOI: 10.17761/1531-2054-26.1.21

[91] Lee DC, Pate RR, Lavie CJ, Sui X, Church TS, Blair SN. Leisure-time running reduces all-cause and cardiovascular mortality risk. Journal of the American College of Cardiology. 2014;**64**(5):472-481. DOI: 10.1016/j.jacc.2014.04.058. Erratum in: J Am Coll Cardiol. 2014 Oct 7;64(14):1537

[92] Dietrich ZC, Joye SW, Garcia JA. Natural medicine: Wilderness experience outcomes for combat veterans. The Journal of Experimental Education. 2015;**38**(4):394-406

[93] Bettmann JE, Prince KC, Ganesh K, Rugo KF, Bryan AO, Bryan CJ, et al. The effect of time outdoors on veterans receiving treatment for PTSD. Journal of Clinical Psychology. 2021;**77**(9):2041-2056. DOI: 10.1002/jclp.23139. Epub 2021 Apr 26

[94] Granero-Jiménez J, López-Rodríguez MM, Dobarrio-Sanz I, Cortés-Rodríguez AE. Influence of physical exercise on psychological well-being of young adults: A quantitative study. International Journal of Environmental Research and Public Health. 2022;**19**(7):4282. DOI: 10.3390/ijerph19074282

[95] John JN, Chukwu CM, Ezeukwu AO, John DO, Okezue OC, Uchenwoke CI. The relationship between self-reported physical activity level, haematological indices, and health-related quality of life of people living with human immunodeficiency virus. HIV & AIDS Review. 2019;**18**(3):176-182. DOI: 10.5114/hivar.2019.88356

[96] Hovland JF, Langeland E, Ness O, Skogvang BO. Experiences with physical activity, health and well-being among young adults with serious mental illness. International Journal of Qualitative Studies on Health and Well-being. 2023;**18**(1):2221911. DOI: 10.1080/17482631.2023.2221911

[97] Ludyga S, Gerber M, Pühse U, Looser VN, Kamijo K. Systematic review and meta-analysis investigating moderators of long-term effects of exercise on cognition in healthy individuals. Nature Human Behaviour. 2020;**4**(6):603-612

[98] Stillman CM, Esteban-Cornejo I, Brown B, Bender CM, Erickson KI. Effects of exercise on brain and cognition across age groups and health states. Trends in Neurosciences. 2020;**43**(7):533-543

[99] Guthold R, Stevens GA, Riley LM, Bull FC. Worldwide trends in insufficient physical activity from 2001 to 2016: A pooled analysis of 358 population-based surveys with 19 million participants. The Lancet Global Health. 2018;**6**(10):e1077-e1086

[100] Pratt M, Varela AR, Salvo D, Kohl HW III, Ding D. Attacking the pandemic of physical inactivity: What is holding us back? British Journal of Sports Medicine. 2020;**54**(13):760-762

[101] Ho V, Merchant RA. The acceptability of digital technology and tele-exercise in the age of COVID-19: Cross-sectional study. JMIR Aging. 2022;**5**(2):e33165

[102] Browne JD, Boland DM, Baum JT, Ikemiya K, Goldman P, Dolezal BA. Lifestyle modification using a wearable biometric ring and guided feedback improve sleep and exercise behaviors: A 12-month randomized, placebo-controlled study. Frontiers in Physiology. 2021;**12**:777874

[103] Sitges C, Terrasa JL, García-Dopico N, Segur-Ferrer J, Velasco-Roldán O, Crespí-Palmer J, et al. An educational and exercise mobile phone–based intervention to elicit electrophysiological changes and to improve psychological functioning in adults with nonspecific chronic low back pain (backfit app): Nonrandomized clinical trial. JMIR mHealth and uHealth. 2022;**10**(3):e29171

[104] Wilke J, Mohr L, Yuki G, Bhundoo AK, Jiménez-Pavón D, Laiño F, et al. Train at home, but not alone: A randomised controlled multicentre trial assessing the effects of live-streamed tele-exercise during COVID-19-related lockdowns. British Journal of Sports Medicine. 2022;**56**(12):667-675

[105] Ozaslan B, Patek SD, Breton MD. Impact of daily physical activity as measured by commonly available wearables on mealtime glucose control in type 1 diabetes. Diabetes Technology & Therapeutics. 2020;**22**(10):742-748

[106] Fabbrizio A, Fucarino A, Cantoia M, De Giorgio A, Garrido ND, Iuliano E, et al. Smart devices for health and wellness applied to tele-exercise: An overview of new trends and technologies such as IoT and AI. Healthcare. 2023;**11**(12):1805

[107] Farley OR, Spencer K, Baudinet L. Virtual reality in sports coaching, skill acquisition and application to surfing: A review. Journal of Human Sport and Exercise. 2019;**15**(3):540-556

[108] Cetin H, Kose N, Gurses E, Turkmen C, Aksoy S, Oge HK. Effects of virtual reality for postural control in chronic neck pain: A single-blind, randomized controlled study. Journal

of Manipulative and Physiological Therapeutics. 2023;**46**(4):191-200

[109] Zeng N, Pope Z, Lee JE, Gao Z. Virtual reality exercise for anxiety and depression: A preliminary review of current research in an emerging field. Journal of Clinical Medicine. 2018;7(3):42

[110] Sheikholeslami S, Ghanbarian A, Azizi F. The impact of physical activity on non-communicable diseases: Findings from 20 years of the Tehran lipid and glucose study. International Journal of Endocrinology and Metabolism. 2018;**16**(4 Suppl):e84740

[111] Romanowska A, Morawiak A, Woods C, Kelly L, Volf K, Gelius P, et al. Health enhancing physical activity policies in Poland: Findings from the HEPA PAT survey. International Journal of Environmental Research and Public Health. 2022;**19**(12):7284

[112] Lee IM, Shiroma EJ, Lobelo F, Puska P, Blair SN, Katzmarzyk PT. Effect of physical inactivity on major non-communicable diseases worldwide: An analysis of burden of disease and life expectancy. The Lancet. 2012;**380**(9838):219-229

[113] Rhodes RE, Janssen I, Bredin SS, Warburton DE, Bauman A. Physical activity: Health impact, prevalence, correlates and interventions. Psychology & Health. 2017;**32**(8):942-975

[114] Snedden TR, Scerpella J, Kliethermes SA, Norman RS, Blyholder L, Sanfilippo J, et al. Sport and physical activity level impacts health-related quality of life among collegiate students. American Journal of Health Promotion. 2019;**33**(5):675-682

[115] Henriksson M, Wall A, Nyberg J, Adiels M, Lundin K, Bergh Y, et al. Effects of exercise on symptoms of anxiety in primary care patients: A randomized controlled trial. Journal of Affective Disorders. 2022;**297**:26-34

[116] Liu Y, Zhang H, Xu R. The impact of technology on promoting physical activities and mental health: A gender-based study. BMC Psychology. 2023;**11**(1):298

Chapter 3

Physical Activity and Sports as a Means to Improve Well-Being

Yusaku Sugiura and Noriko Higuchi

Abstract

Physical inactivity has emerged as a global health concern, contributing to an increase in lifestyle-related diseases and a decline in healthy life expectancy. In Japan, where life expectancy is among the highest in the world, promoting physical activity habits at all stages of life is a pressing issue. This study explores strategies to promote physical activity and well-being, focusing on Japanese university students. Using the seesaw model of behavior change, which balances motivation and strain with self-determination as the fulcrum, we examined educational interventions using health support applications, social networking sites, reflective dialog, and self-directed learning. Results showed that short-term use of exercise support applications increased exercise self-efficacy among university students but did not significantly increase physical activity levels. Furthermore, combining technology with peer interaction and reflective dialog facilitated students' awareness, agency, and behavior change in relation to physical activity. We propose a paradigm shift in physical education from skill acquisition to experiential learning based on the TAIIKU (Physical Education to Sports Continuum) framework to cultivate lifelong wellness literacy and autonomous engagement in physical activity in response to contemporary societal challenges such as the volatile, uncertain, complex, ambiguous (VUCA) era.

Keywords: physical activity, physical education, sports, wellness, well-being

1. Introduction

In 2018, the World Health Organization (WHO) reported that more than 1 billion people are inactive, corresponding to one in three adults globally [1]. Physical inactivity worldwide has surged, resembling a pandemic with few signs of improvement [1]. If exercise habits are not established, the rate of lifestyle-related diseases may increase, eventually reducing the population's healthy life expectancy. This also increases medical costs and reduces the labor force, leading to a loss of vitality in society. The economic impact of declining exercise habits is estimated to be \$53–\$58 billion globally [2].

To motivate people to adopt regular physical exercise, it is essential to clarify its significance and to learn how to practice it effectively. Cultivating a sense of well-being during university years is also beneficial because it can significantly impact the development of exercise habits later in life [3]. Additionally, establishing exercise habits can be tailored to meet the needs of each individual. This argument has been supported by previous research on health education interventions that promote physical activity and well-being. Fostering well-being can encourage more people to engage in physical activity, potentially enhancing societal productivity [3].

Ryff and Keyes [4] found that psychological well-being goes beyond happiness; it comprises six necessary elements for well-being in life: (1) self-acceptance, (2) positive relations with others, (3) autonomy, (4) environmental mastery, (5) purpose in life, and (6) personal growth. Durden-Myers et al. [5] added optimism and vitality (as components of human flourishing) to the list, demonstrating their relationship with physical literacy. Meanwhile, physical education that embraces the concept of physical literacy can lead to self-actualization, greater resilience, and improved social relationships [6] and healthy life expectancy [7].

Physical education in Japan and abroad is increasingly being re-examined, as the future of physical education must focus on physical literacy and activity to enrich people's lives and promote wellness in their lifestyles [8]. In today's society, characterized by abundant goods and technological advancements, daily life has become highly convenient, fostering an environment where individuals may not necessarily perceive the need to exercise. In this context, the concept of physical literacy and the broader field of sports and physical education have been gaining attention in Japan.

To advance wellness through sports and physical education in the future, it is essential to develop individualized "life designs" that enable individuals to comprehend the importance of physical activity. To cultivate the ability to design a fulfilling life, the following four elements are necessary: (1) awareness and knowledge of health, (2) basic physical abilities necessary for exercise, (3) motivation to be physically active and healthy, and (4) lifelong engagement with physical activity. These components are typically acquired through physical literacy education, which is crucial for improving individual capabilities in these areas. We are responsible for understanding and promoting the value of motivation, confidence, and physical activity throughout our lives. Physical activity promotes athletic performance and fosters lifelong enjoyment, ultimately contributing to overall well-being.

Since the COVID-19 pandemic, our society has been radically transformed, with increased awareness of the convenience of online systems. For example, due to the efficiency of modern work practices, the time spent commuting to work can now be adjusted according to one's needs. At the same time, leisure time created by technological advancements has opened up opportunities for physical exercise.

Finally, as information technology and artificial intelligence expand, we are entering the so-called VUCA (volatile, uncertain, complex, ambiguous) era, prompting individuals to reconsider how to improve their well-being. The cornerstone of this improvement is wellness through physical activity. Because physical activity during university years can significantly impact the development of exercise habits in later life, this study examined the importance of physical activity and sports in wellness education among a sample of Japanese university students.

2. Seesaw model theory of practicing physical activity

Among the factors that positively influence well-being, self-determination enables individuals to intentionally shape their behavior [9]. In this context, a person's behavior is influenced by the balance between "burden" (or psychological load) and "motivation," with the latter being the facilitator and the former being the inhibitor. This behavioral practice can be referred to as a "seesaw," as illustrated in the seesaw model [10]. According to this theory, the seesaw tilts toward motivation when motivation surpasses the burden, leading to behavioral practice. This balance depends on the position of the seesaw's fulcrum, which corresponds to the individual's capacity for self-determination.

To enhance motivation and sustain physical activity, one or more of the following factors must be achieved: (1) enhanced motivation, (2) reduced burden, (3) improved self-determination, and (4) more emotional support that promotes self-determination. The interconnectedness of these factors makes physical activity more sustainable (**Figure 1**) [11].

In general, strengthening the motivation for physical activity involves clarifying the reasons for engaging in such activity and understanding its benefits. This process is also closely linked to fostering a sense of purpose in life. Conversely, reducing the burden associated with physical activity helps minimize one's aversion to it. In addition to improving self-determination, building confidence in one's ability to engage in and sustain physical activity is important. Meanwhile, promoting and maintaining physical activity requires making independent choices based on personal interests and building peer relationships. Cultivating a sense of self-efficacy through various challenges and constructive feedback is also important. Interventions based on the self-determination theory indicate that physical activity is more likely to be sustained when individuals increase their intrinsic motivation [12, 13]. Specific methods for achieving this include positive self-talk (i.e., encouraging oneself with uplifting words) and self-counseling, which involves organizing thoughts to reduce anxiety about one's actions.

However, even when individuals want to change their behavior, they do not always act rationally. Additionally, psychological conflicts, such as cravings, can emerge, making self-determination even more challenging [14]. Hence, emotional support through safe, open dialog can effectively promote self-determination despite these

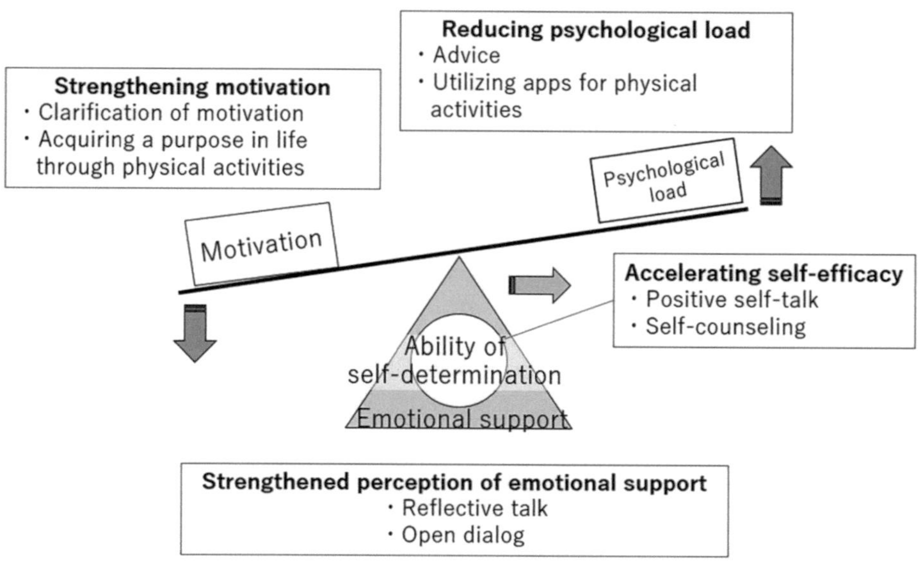

Figure 1.
Support method for promoting physical activities based on seesaw model [11].

conflicting desires. Engaging in reflective conversations allows individuals to share specific experiences with their peers, thus deepening their self-understanding and strengthening their self-determination. Regular physical activity positively affects mental health by reducing stress and anxiety while increasing self-esteem and overall well-being [15–18].

3. Use of technology to enhance motivation, reduce psychological burden, and promote self-efficacy

3.1 Impact of health and exercise support applications on physical activity

Previous research has found that the physical activity of first-year university students tends to decline [19]. At the same time, many applications that support health and exercise have been developed for use through mobile devices, increasing research on physical activity utilizing these applications [20–23]. According to the seesaw model, such applications can enhance motivation, reduce perceived burdens, and promote self-efficacy by visualizing exercise habits. Support for enhancing motivation can be classified into two categories: (1) "behavioral approaches," such as goal setting, contracting, self-monitoring, cues, and rewards; and (2) "cognitive approaches," including decision-making, health education, and information provision. The former is designed to increase competence in physical activity and tends to be more effective than the latter [23].

In Japan, limited empirical research has been conducted on the prevalence of health and exercise support applications and their impact on physical activity levels. To address this gap, we examined the effects of intervention applications among a sample of first-year university students in prior research [24]. Specifically, a questionnaire survey was conducted with 273 first-year university students (133 males; 140 females) to evaluate their stages of change in exercise behavior, frequency, enjoyment, and self-efficacy, as well as their use of such applications.

The results for the stages of change were as follows: 31.5% in the maintenance phase, 7.3% in the indifference phase, 16.5% in the interest phase, 31.5% in the preparation phase, and 13.2% in the execution phase. Those in the maintenance and execution phases had significantly higher favorability for exercise self-efficacy and confidence in their health behaviors ($p < .001$). Additionally, 251 participants (92%) used health and exercise support applications. The duration of use was 60.6% for 1 week or less, 25.0% for 2–4 weeks, 6.4% for 5–7 weeks, and 8.0% for 8 weeks. Meanwhile, exercise frequency (days/week) was 2.24 ± 1.9 before the intervention. However, exercise frequency significantly decreased to 2.09 ± 1.9 after 2 months ($p < .05$). Moreover, those who utilized the application for 4 weeks or less showed a significant decrease in exercise frequency, whereas those who used it continuously for 5 weeks or more did not. These results indicate that the continuous use of health and exercise support applications during the early university years can effectively mitigate the decline in physical activity.

3.2 Impact of combining health and exercise support applications and mobile social networking site (SNS) applications

Health and exercise support applications effectively promote physical activity; however, they do so mainly in the short term [25]. However, using these applications with other approaches can lead to long-term behavioral changes and enhanced effectiveness [25]. Integrating mobile social networking site (SNS) applications with health and exercise support applications allows users to share their activities [26]. This sense of community can foster regular exercise habits. At the same time, using social magnetism in mobile SNS applications can positively influence exercise promotion by improving individual health and increasing familiarity [27, 28].

Consequently, in previous research, we examined how integrating a health and exercise support application with a mobile SNS application affects university students who do not have established exercise habits. The study focused on physical activity (measured in mileage and frequency) and exercise self-efficacy [29]. A total of 19 university students participated in a 16-week intervention, after which data from the final 16 participants were analyzed. The intervention comprised two phases: Phase I (weeks 1–8) involved using an exercise support application to visualize exercise history, while Phase II (weeks 8–16) integrated this application with a mobile SNS application, providing opportunities for participants to interact with their peers and share their exercise history. In this case, the participants engaged in various activities, including relay races and trail runs.

Conversely, the results regarding the distance traveled or the number of runs showed no significant changes after Phases I and II. However, exercise self-efficacy increased after Phase I ($p = 0.006$). Additionally, there was a positive correlation between the number of posts on the SNS and distance traveled ($\rho = 0.545$, $p = 0.036$). This correlation was based on several factors: visualizing their history, which increased their confidence; being motivated by peers' running posts; enjoying running more; and having more opportunities to invite their peers to join. However, the lack of face-to-name recognition made users hesitant to post on SNS.

In summary, combining an exercise support application and a mobile SNS application enhanced exercise self-efficacy among university students. However, the increase in the amount of exercise was not confirmed. Thus, to increase exercise levels, a mechanism must be developed that fosters a sense of unity among users and strengthens connections with peers. By carefully forming user groups, integrating exercise support applications with mobile SNS applications can catalyze the development of exercise habits among this demographic.

4. Need for reflective dialog to promote self-efficacy and provide support

In late 2019 and early 2020, COVID-19 spread throughout the world, limiting people's ability to interact freely. Before the pandemic, physical activity was promoted through various applications. However, after January 2020, physical activity significantly declined, especially among university students [30, 31]. In this context, we reaffirm the importance of face-to-face gatherings, direct dialog, and interactions, believing that these connections are essential for nurturing agency and developing healthy exercise habits.

4.1 Using reflective dialog to routinize physical activity

Reflective dialog has been used to promote agency and improve understanding and connection among individuals [32]. We incorporated reflecting talk which is a part of the speaking and listening dialog approach. This talk can be divided into careful external and internal conversations [33]. These reflections can potentially foster agency in university students, among others.

The present study examined how reflective dialog can increase interest in physical activity and help individuals establish healthy exercise habits [34]. Specifically, 13 university students (12 males and 1 female) participated in three sports-related dialogs. Each 60-minute dialog was conducted in groups of three or four students. Each session included a reflective component in which the participants discussed the importance of physical activity and how to maintain it in their daily lives. Meanwhile, the changes in the students' awareness and motivation toward physical activity throughout the dialogs were analyzed using the Steps for Coding and Theorization (SCAT) framework. The findings are as follows.

4.1.1 The 1st dialog

In this dialog, the university students acknowledged the importance of maintaining their motivation for physical activity. Collectively, they explored ways to become more active by leveraging mobile devices and applications. Overall, they integrated physical activity into their daily lives by setting small, achievable goals.

4.1.2 The 2nd dialog

In this dialog, the university students recognized the importance of engaging in physical activities and visualizing specific exercises. This process fostered greater self-awareness of their physical activity. The students also gained insights into the advantages/disadvantages of mandatory goal setting and acknowledged the impact of group dynamics and motivation through various experiences.

4.1.3 The 3rd dialog

In this dialog, the university students recognized the enjoyment of communication and discussed the advantages and disadvantages of shared goals and outcomes. The participants also compared competitive and self-reward motivation and discussed various issues related to healthy exercise habits.

These reflective dialogs facilitated discussions about motivation for physical activity among the participating university students. The findings indicate that enjoyment and technology can help maintain this motivation, whereas self-awareness and group dynamics can positively influence behavioral changes related to physical activity. However, it is possible that the dialogs primarily influenced participants' awareness and perceptions. Therefore, future efforts should incorporate a practical stage combining reflective dialog with physical activity to evaluate actual behavioral changes.

4.2 Influence of self-directed learning on sports and physical activity

In this study, we also investigated how self-directed learning influences the perceptions and behaviors of university students regarding physical activity [35]. Specifically, 30 university students (22 males and 8 females) participated in a 12-week program involving physical activities, fieldwork, and sports-related discussions initiated by the students. During these dialogs, the participants formed small groups of three or four individuals and shared their views and insights on physical activity. These reflections were then analyzed using the SCAT framework [36].

The students' self-efficacy scores for exercise increased slightly after the intervention. After the fieldwork, the students recognized changes in their perspectives on

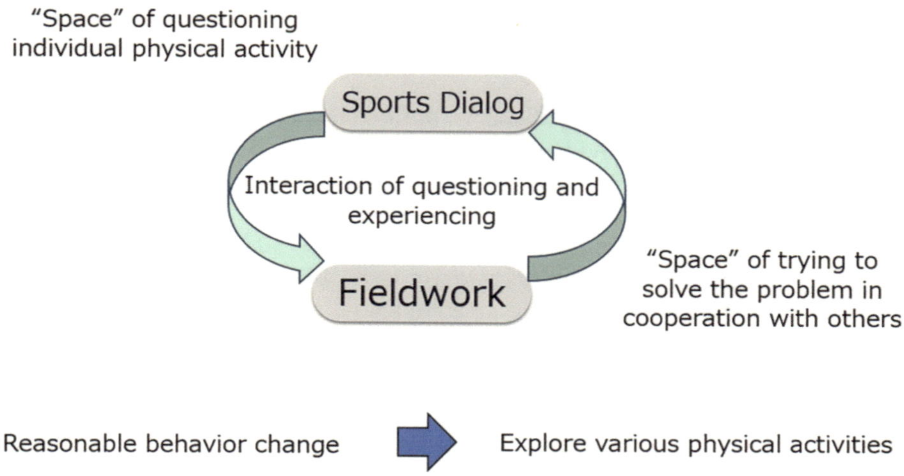

Figure 2.
"Space" for self-directed learning to make sports an everyday activity.

health, particularly regarding physical activity. Additionally, through the dialogs, they became more aware of several factors that inhibit exercise habits, broadening their understanding of wellness and the dynamics of physical activity (**Figure 2**).

Based on these findings, this self-directed learning initiative established a physical activity routine and offered the students a lifelong perspective on the importance of such activity [35]. Although the program did not significantly increase overall physical activity, it addressed the students' needs and objectives, promoting their autonomous motivation [37].

Finally, the importance of "space" where people can gather and work was reaffirmed. In this regard, a "space" is a physical and relational space where knowledge is generated, shared, and developed [38].

A space for sharing experiences through dialog is essential for transforming individual tacit knowledge into collective knowledge [39]. In this regard, self-directed learning can be an effective platform for providing emotional support, encouraging others' opinions, fostering deeper reflection, pursuing new goals with peers, and promoting confidence in physical activity. Practice and research on exercise habits have revealed that collaboration with others is essential for fostering general well-being.

5. Physical education and sports for promoting wellness

5.1 Re-examining the significance of physical education and sports in promoting wellness

"Exercise" refers to planned, deliberate, and sustained physical activities aimed at maintaining and improving physical fitness. Conversely, "daily life activities" encompass the physical activities involved in everyday life, such as employment, housework, and commuting.

The Japan Sports Agency has launched various initiatives to promote health through sports, including "Sport in Life" [40]. This program makes sports more accessible and encourages the public to establish regular exercise routines. It also broadens the definition of "sports" to include various physical activities, lowering participation barriers. The goal is to facilitate behavioral changes in daily activities (**Figure 3**).

In Japan, the original meaning of sports, defined by play and enjoyment, has largely diminished. Many individuals now equate "sports" with physically demanding activities, perceiving it as being in athletic apparel and engaging in intense training regimens. Nevertheless, various initiatives have led to a gradual increase in the population's participation in regular exercise—defined as 30 minutes of activity on two or more days per week—rising to approximately 50% [41]. This indicates that physical fitness should be promoted as an enjoyable activity in Japan and elsewhere. Fostering wellness based on each individual's subjective view of health, instead of being constrained by traditional healthcare frameworks that focus on treatment and prevention, is important.

5.2 The TAIIKU (physical education to sports) system for wellness

In recent years, there has been a growing shift from physical education to sports, both in Japan and abroad. However, we argue against this trend of diminishing the importance of physical education. In Japan, sports—whether competitive or health-oriented—have a rich history and are often embedded within physical education programs, reflecting a broader culture of wellness. To that end, we should further promote wellness through activities such as judo, which originated in Japan, and showcase the strengths of the Japanese. Meanwhile, there is a need to reassess the value of physical education within the continuum of "physical education to sports" (**Figure 4**) [42, 43] . "TAIIKU," which refers to physical education, is a concept that describes all human physical activities [43] and is a broad concept of physical education.

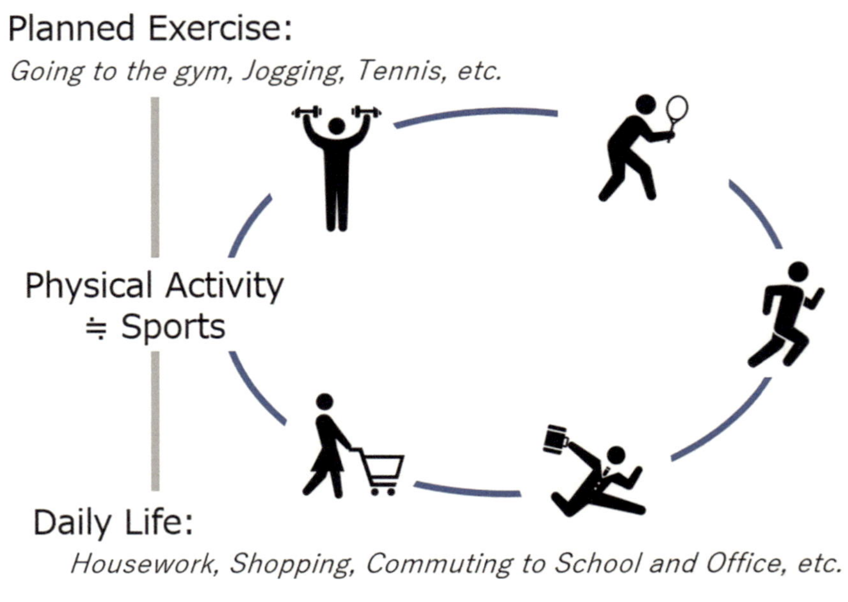

Figure 3.
"Sports in life".

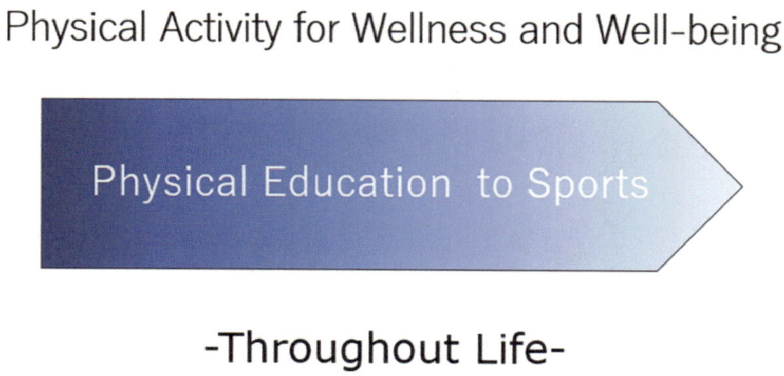

Figure 4.
TAIIKU (physical education to sports) system.

Physical education is essential for developing a strong body, both physically and mentally, as it provides various stimuli through exercise. However, physical education should extend beyond simply educating individuals about the physical body. Physical education should not focus solely on the "education of the physical fitness." The original purpose of physical education is the "education through the physical (activity)" to improve fitness and wellness [44]. In Japan, qualities such as "civility" (respect) and "cooperativeness," fostered through the unique culture of physical education, have become defining characteristics of the Japanese people. This indicates that physical education should emphasize acquiring the basic fitness needed for daily life, encouraging healthy lifestyles, and developing non-cognitive skills. Meanwhile, the essence of sports lies in the enjoyment and comfort derived from physical exercise. Thus, we must preserve the educational value of the TAIIKU system, a continuum of physical education to sports, as a vital aspect of promoting wellness throughout life.

5.3 Paradigm shift to promote wellness through the TAIIKU system

Physical education methods have primarily focused on developing practical skills in exercise and sports, aiming to promote health and physical fitness in Japan and around the world. However, fostering exercise habits should not only focus on practical skills but also emphasize behavioral changes and awareness through collaboration with others.

Building on Kolb's learning theory, which comprises the four components of diverging, assimilating, converging, and accommodating, Korthagen et al. [45] proposed the ALACT (action, looking back, awareness of essential aspects, creating alternative methods of action, trial) model. This model outlines individuals' learning cycles as follows: (1) perform an action; (2) reflect on the action; (3) become aware of essential aspects; (4) expand their options for the action; and (5) make new attempts (**Figure 5**).

Generally, physical education and sports are defined by three characteristics: physicality, simultaneity, and collaboration. These characteristics are closely related to daily physical activity, including exercise and daily activities. It is well established that regular physical activity is essential for maintaining physical and mental health. However, the challenge is to shift the focus from traditional "expert"-led education to a model in which individuals actively engage with one another. In this case, this shift

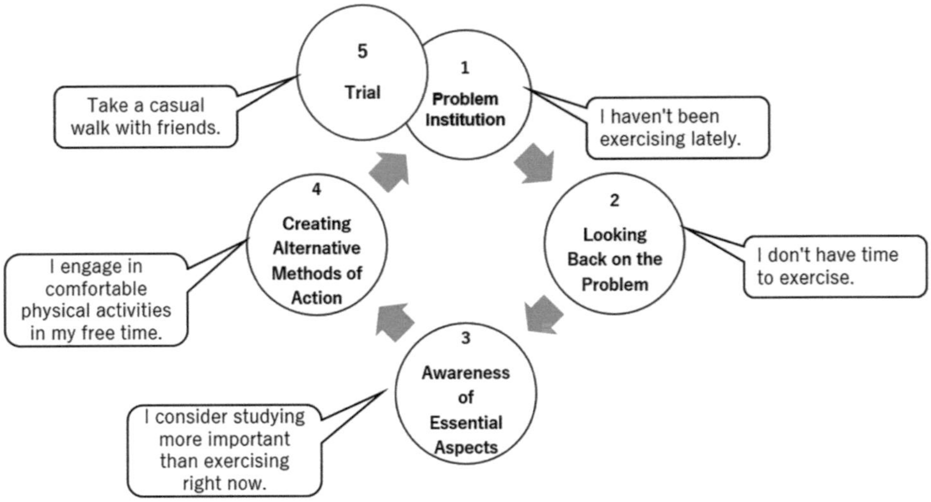

Figure 5.
An example of a reflective cycle for physical activity.

	Physical education/sports	
	Previous	**Future**
Purpose	Health and physical fitness	Wellness and well-being
Leading position	Teacher/Instructor	Collaborative facilitator
View of education	Seek answers	Ask questions
Leaning style	Instruction/guidance	Self-directed learning
Communication style	Unidirectional	Bidirectional (Dialog)
Motivation	Extrinsic	Intrinsic
Assessment	Physical	Wellness

Table 1.
Paradigm shift through TAIIKU system.

should include employing the TAIIKU system to foster wellness and empower individuals to take the initiative in physical activity.

In the future, especially in a society with a 100-year life expectancy and given the challenges of the VUCA era, reliance on previous successes and mere knowledge acquisition will no longer suffice. Education must embrace learning through self-initiated questioning, reflection, and ongoing dialog with others. Therefore, the TAIIKU system holds the potential to strengthen life management skills and contribute to holistic well-being (**Table 1**).

6. Conclusion

In Japan, there has been a longstanding TAIIKU system for sports and physical education. However, this system has not been effective because of insufficient support. Furthermore, educators and practitioners often do not fully understand how physical education and sports can promote wellness. They tend to force students to participate in activities rather than foster independence. As a result, various issues have emerged, such as "not understanding the significance of physical activity," "not understanding how to exercise," and "not enjoying exercise," especially among university students without a strong foundation in sports [3, 46]. Although various strategies have been implemented to encourage exercise habits, their effectiveness remains questionable [3]. Therefore, we recommend an educational approach grounded in experiential learning that can help develop a physical education and sports framework to enhance physical literacy, promote well-being, and connect individuals with peers through active participation in physical activities.

Acknowledgements

This work was supported by a JSPS KAKENHI Grant (No. JP20K11449).

Conflict of interest

The authors declare no conflict of interest.

Author details

Yusaku Sugiura* and Noriko Higuchi
Meikai University, Urayasu, Japan

*Address all correspondence to: yusaku@meikai.ac.jp

References

[1] World Health Organization. Nearly 1.8 Billion Adults at Risk of Disease from Not Doing Enough Physical Activity. 2024. Available from: https://www.who.int/news/item/26-06-2024-nearly-1.8-billion-adults-at-risk-of-disease-from-not-doing-enough-physical-activity [Accessed: April 20, 2025]

[2] Ding D, Lawson KD, Kolbe-Alexander TL, Finkelstein EA, Katzmarzyk PT, van Mechelen W, et al. Lancet physical activity series 2 executive committee. The economic burden of physical inactivity: A global analysis of major non-communicable diseases. Lancet. 2016;**388**:1311-1324. DOI: 10.1016/S0140-6736(16)30383-X

[3] Sugiura Y, Higuchi N. Considering the significance of physical activity in an age of uncertainty: A proposal for the introduction of physical. Journal of the Japan Academy for Health Behavioral Science. 2021;**35**:15-22. In Japanese

[4] Ryff CD, Keyes CLM. The structure of psychological well-being revisited. Journal of Personality and Social Psychology. 1995;**69**:719-727. DOI: 10.1037//0022-3514.69.4.719

[5] Durden-Myers EJ, Whitehead ME, Pot N. Physical literacy and human flourishing. Journal of Teaching in Physical Education. 2018;**37**:308-311. DOI: 10.1123/jtpe.2018-0132

[6] Whitehead M. Physical Literacy across the World. London: Routledge; 2019

[7] Whitehead M. Definition of physical literacy and clarification of related. ICSSPE Bulletin Journal of Sport Science and Physical Education. 2013;**65**:28-33

[8] Edwards LC, Bryant AS, Keegan RJ, Morgan K, Jones AM. Definitions, foundations and associations of physical literacy: A systematic review. Sports Medicine. 2017;**47**:113-126. DOI: 10.1007/s40279-016-0560-7

[9] Martela F, Lehmus-Sun A, Parker PD, Pessi AB, Ryan RM. Needs and well-being across Europe: Basic psychological needs are closely connected with well-being, meaning, and symptoms of depression in 27 European countries. Social Psychological and Personality Science. 2023;**14**:501-514. DOI: 10.1177/19485506221113678

[10] Munakata T. Psycho-social influence on self-care of the hemodialysis patient. Social Science and Medicine. 1982;**16**:1253-1264. DOI: 10.1016/0277-9536(82)90069-7

[11] Higuchi N, Sugiura Y. Promoting physical activity among female college students using mobile applications. Journal of the Japan Academy for Health Behavioral Science. 2020;**35**:22-29. In Japanese

[12] Ryan RM, Deci EL. Self-Determination Theory: Basic Psychological Needs in Motivation, Development, and Wellness. New York: Guilford Press; 2019

[13] Fortier MS, Sweet SN, O'Sullivan TL, Williams GC. A self-determination process model of physical activity adoption in the context of a randomized controlled trial. Psychology of Sport and Exercise. 2007;**8**:741-757. DOI: 10.1016/j.psychsport.2006.10.006

[14] Hashimoto S, Suzuki k, Komori M. A qualitative study of outward and inward intentions based on the double-intention model. Journal of the Japan Academy for Health Behavioral Science. 2003;**18**:122-139. In Japanese

[15] Hills AP, Dengel DR, Lubans DR. Supporting public health priorities: Recommendations for physical education and physical activity promotion in schools. Progress in Cardiovascular Diseases. 2015;**57**:368-374. DOI: 10.1016/j.pcad.2014.09.010

[16] Biddle S. Physical activity and mental health: Evidence is growing. World Psychiatry. 2016;**15**:176-177. DOI: 10.1002/wps.20331

[17] Chekroud SR, Gueorguieva R, Zheutlin AB, Paulus M, Krumholz HM, Krystal JH, et al. Association between physical exercise and mental health in 1·2 million individuals in the USA between 2011 and 2015: A cross-sectional study. The Lancet Psychiatry. 2018;**5**:739-746. DOI: 10.1016/S2215-0366(18)30227-X

[18] Buecker S, Simacek T, Ingwersen B, Terwiel S, Simonsmeier BA. Physical activity and subjective well-being in healthy individuals: A meta-analytic review. Health Psychology Review. 2021;**15**:574-592. DOI: 10.1080/17437199.2020.1760728. Google Scholar

[19] Thomas AM, Beaudry KM, Gammage KL, Klentrou P, Josse AR. Physical activity, sport participation, and perceived barriers to engagement in first-year Canadian university students. Journal of Physical Activity and Health. 2019;**16**:437-446. DOI: 10.1123/jpah.2018-0198

[20] Modave F, Bian J, Leavitt T, Bromwell J, Harris C, Vincent H. Low quality of free coaching apps with respect to the American college of sports medicine guidelines: A review of current mobile apps. JMIR mHealth and uHealth. 2015;**3**:e77. DOI: 10.2196/mhealth.4669

[21] Payne HE, Moxley VB, MacDonald E. Health behavior theory in physical activity game apps: A content analysis. JMIR Serious Games. 2015;**3**:e4. DOI: 10.2196/games.4187

[22] Romeo A, Edney S, Plotnikoff R, Curtis R, Ryan J, Sanders I, et al. Can smartphone apps increase physical activity? Systematic review and meta-analysis. Journal of Medical Internet Research. 2019;**21**:e12053. DOI: 10.2196/12053

[23] Conn VS, Hafdahl AR, Mehr DR. Interventions to increase physical activity among healthy adults: Meta-analysis of outcomes. American Journal of Public Health. 2011;**101**:751-758. DOI: 10.2105/AJPH.2010.194381

[24] Higuchi N, Sugiura Y. Effect of health and exercise applications on the level of physical activities in female university students at the earliest stage of college life: Survey on gender differences. The Journal of Japan Mibyo System Association. 2018;**24**:47-51. In Japanese

[25] Direito A, Jiang Y, Whittaker R, Maddison R. Apps for IMproving FITness and increasing physical activity among young people: The AIMFIT pragmatic randomized controlled trial. Journal of Medical Internet Research. 2015;**17**:e210. DOI: 10.2196/jmir.4568

[26] Knight E, Stuckey MI, Prapavessis H, Petrella RJ. Public health guidelines for physical activity: Is there an app for that? A review of android and apple app stores. JMIR mHealth and uHealth. 2015;**3**:32. DOI: 10.2196/mhealth.4003

[27] Schweier R, Romppel M, Richter C, Hoberg E, Hahmann H, Scherwinski I, et al. A web-based peer-modeling intervention aimed at lifestyle changes in patients with coronary heart disease and chronic back pain: Sequential controlled trial. Journal of Medical Internet

Research. 2014;**16**:e177. DOI: 10.2196/jmir.3434

[28] Centola D. The spread of behavior in an online social network experiment. Science. 2010;**329**:1194-1197. DOI: 10.1126/science.1185231

[29] Sugiura Y, Higuchi N, Hoshino N. Impacts of use of exercise assistance and SNS mobile applications on exercise quantity and self-efficacy of exercise habit in university students. The Journal of Japan Mibyo System Association. 2017;**23**:30-34. In Japanese

[30] Romero-Blanco C, Rodríguez-Almagro J, Onieva-Zafra MD, Parra-Fernández ML, Prado-Laguna MDC, Hernández-Martínez A. Physical activity and sedentary lifestyle in university students: Changes during confinement due to the COVID-19 pandemic. International Journal of Environmental Research and Public Health. 2020;**17**:6567. DOI: 10.3390/ijerph17186567

[31] López-Valenciano A, Suárez-Iglesias D, Sanchez-Lastra MA, Ayán C. Impact of COVID-19 pandemic on university students' physical activity levels: An early systematic review. Frontiers in Psychology. 2020;**11**:624567. DOI: 10.3389/fpsyg.2020.624567

[32] Kawai N. Positive reflective mindsets emerge in a community of learners through dialogue-based higher education practices: A retrospective interview of students attending dialogical seminars. In: Proceedings of the Research Institute of Sapporo Gakuin University. Vol. 10. 2023. pp. 23-34. In Japanese

[33] Andersen T. The Reflecting Team Dialogue and Dialogues about the Dialogues. New York: WW Norton & Co., Inc.; 1991

[34] Higuchi N. An attempt at 'sports dialogue' to make physical activity a part of daily life. Meikai Journal, Faculty of Languages and Cultures. 2024;**36**:51-64. In Japanese

[35] Otani T. Steps for coding and theorization, SCAT a qualitative data analysis method by four-step coding: Easy. Startable and small scale data. Educational Sciences. 2008;**54**:27-44. In Japanese

[36] Higuchi N, Kobayashi Y, Hasui T, Sugiura Y. Collaborative activities that make sports routine. Journal of the Japan Academy for Health Behavioral Science. 2024;**39**:41. In Japanese

[37] Goncalves A, Deshayes M, Gisclard B, Philippe AG, Bernal C, Krawczyk S, et al. Exploring the health effectiveness of a physical activity program co-constructed with students after the COVID-19 pandemic. Nutrients. 2023;**15**:2913. DOI: 10.3390/nu15132913

[38] Nonaka I, Konno N. The concept of 'Ba': Building a foundation for knowledge creation. California Management Review. 1998;**40**:40-54. DOI: 10.2307/41165942

[39] Nonaka I, Toyama R, Konno N. SECI, Ba and leadership: A unified model of dynamic knowledge creation. Long Range Planning. 2000;**33**:5-34. DOI: 10.1016/S0024-6301(99)00115-6

[40] Japan Sports Agency. Sport in Life purojekuto. 2024. In Japanese. Available from: https://www.mext.go.jp/sports/b_menu/sports/mcatetop05/list/1396542_00001.htm [Accessed: April 20, 2025]

[41] Japan Sports Agency. Reiwa 4 nendo "Supo-tsu no Jisshi Joukyou Nado ni Kansuru Yoron Chosa" no Kekka ni tsuite. 2023. In Japanese. Available from: https://www.mext.go.jp/sports/b_menu/houdou/jsa_00133.html [Accessed: April 20, 2025]

[42] Ae M. Supotsu Cho, Supo-tsu Shingikai (Supo-tu Kihon Keikai Bukai) Iken Ukagai no Kai Memo Sho. 2016. In Japanese. Available from: https://www.mext.go.jp/sports/b_menu/shingi/001_index/bunkabukai/shiryo/__icsFiles/afieldfile/2016/08/05/1375312_004.pdf [Accessed: April 20, 2025]

[43] Yosuke H. Taiiku no Gaku towa nanika, Dowa Shoin, Tokyo. 2023. pp. 21-31

[44] Williams JF. Education through the physical. The Journal of Higher Education. 1930;**1**:279-282

[45] Korthagen FAJ, Kessels J, Koster B, Lagerwerf B, Wubbels T. Linking Practice and Theory: The Pedagogy of Realistic Teacher Education. London: Routledge; 2001. DOI: 10.4324/9781410600523

[46] Hyadain, Tanomukara Sotto Shiteoite Kudasai. Taiiku ka Kyouiku. Vol. 3. Tokyo: Taishukan Publishing Co., Ltd.; 2019. p. 9. In Japanese

Chapter 4

Physical Activity and Sports as Preventive Medicine for Psychosocial Health and Well-Being for Adolescents

Nasser Abouzeid and Abdullah Aldughaysh

Abstract

This book chapter examines the significant role of physical activity in enhancing adolescent mental health and overall well-being. It introduces a framework that emphasizes the interconnected nature of physical, cognitive, emotional, social, and autonomous development, suggesting that physical activity can be strategically used to address multiple challenges during adolescence. The chapter highlights that team sports and recreational physical activity are particularly effective in promoting social skills and self-esteem. There are gender differences in the benefits of physical activity, with males often showing more pronounced mental health improvements. The importance of family and social support is also noted in encouraging physical activity and contributing to better mental health outcomes. Furthermore, this chapter explores the integration of physical activity into academic settings to boost cognitive function. It notes the influence of sleep quality and family support on the relationship between physical activity and mental health. This work proposes a model for implementing targeted physical activity programs in various settings to build resilience and self-esteem. The proposed model suggests that programs should integrate cognitive, emotional, social, autonomous, and physical aspects of development. The chapter concludes by recommending future research on the long-term effects of physical activity and the development of specific intervention strategies.

Keywords: physical activity, adolescent mental health, Integrated Developmental Model (IDM), team sports, family and social support, cognitive development, gender differences, intervention programs, recreational physical activity, emotional regulation, psychosocial health, quality of life, high-intensity interval training (HIIT), normative data, lifetime mental disorders, resilience, school-based interventions, community-based programs

1. Introduction

Since the time of early humanity, humans have been doing physical activity out of necessity. Farming and hunting were physically demanding. However, it can be argued that humans' realization of the importance of physical activity was not a recent development, as evidenced by the conception of recreational physical activity. It can be indicated that the recorded history of competitive sports began in the city of Olympia in Greece in the eighth century BC [1]. This shows that humans understand the link between physical activity and overall health. They knew that physical activity increased strength and fitness for military personnel [2]. The link between competitive sports and military readiness has continued into modern times through military sports tournaments [3].

As for mental health, ancient Greek philosophers were also instrumental in clarifying the role of physical activity in mental health through the concept of mind-body balance [4]. The modern era has seen a clear and continuous increase in the popular realization of the importance of physical activity, both physically and psychologically. This has been the result of increasing studies pointing to its benefits. Physical activity can provide similar or even greater benefits than drugs, as shown in Fiuza-Luces et al. [5]. In this study, physical activity illustrated effects in reducing risk factors such as glucose intolerance, blood lipids, blood pressure, and thrombosis. The authors of this study present epidemiological evidence showing the protective effects of exercise and how it can favorably impact the expression of our genome shaped by evolution to support physical activity. Overall, this study makes a strong case for exercise as the "real polypill" that is more accessible, cost-effective, and safer than pharmaceutical interventions [5].

2. Methodology

A basic quantitative synthesis was performed after systematically reviewing and extracting certain numerical data from 23 studies examining the effects of physical activity (PA) on the physiological level as well as the psychological/mental health (MH) level. Data were organized by study, type of PA measure, type of MH measure, and statistical results. The extraction process focused on specific numerical values. These values are identified below along with one or two elected studies representing the numerical value or a statistical technique employed in the reviewed studies.

2.1 Means and standard deviations

Abouzeid et al. [6] provided blood oxygen saturation (SpO_2) levels, with elevation training mask (ETM) users showing a SpO_2 of 94% compared to 96% in the control group. These values were extracted to compare the physiological impacts of different training methods. Saanijoki [7] provides means and standard deviations for various physical measures such as BMI and VO_2 peak for different groups.

2.2 Path coefficients

Menhas et al. [8] reported path coefficients from structural equation modeling (SEM), including the relationship between physical activity and mental health. For example, the path coefficient between physical activity and psychological resilience was 0.631, and the path coefficient between physical activity and quality of life was 0.601. These values were used to assess the strength of relationships within their model.

2.3 Regression coefficients

Liu et al. [9] presented regression coefficients to quantify the association between PA and MH outcomes. They reported that high-frequency PA had a negative association with mental health of −0.439, medium-frequency PA had an association of −0.339, and low-frequency PA had an association of −0.171. These coefficients were extracted to determine the magnitude and direction of these associations.

2.4 Odds ratios (ORs)

Liu et al. [9] reported odds ratios, such as the finding that males meeting PA guidelines were less likely to have depression symptoms (OR = 0.73). These ORs were extracted to compare the likelihood of mental health outcomes based on PA levels.

2.5 Correlation coefficients

Correlation coefficients seen in the study by Kyan et al. [10] were extracted to measure relationships between different variables. For elementary and junior high school pupils, the correlation coefficients between patient-centered assessment and counseling for exercise plus nutrition (PACE+) and accelerometry were 0.13 and 0.50, respectively.

2.6 Direction of effects

Abouzeid et al.'s [6] data showed that ETM use led to lower blood oxygen saturation (SpO_2) levels compared to a control group (94% vs. 96%, respectively), indicating a physiological impact of the intervention. Menhas et al. [8] reported positive path coefficients between physical activity and mental health (0.631 and 0.601), psychological resilience (0.631 and 0.592), and quality of life (0.601 and 0.592), suggesting that greater physical activity was associated with positive mental health outcomes. Liu et al. [9] showed negative regression coefficients between physical activity frequency and mental health, demonstrating that higher physical activity frequency is associated with better mental health.

2.7 Magnitude of effects

Liu et al. [9] showed that the negative association with mental health increased with physical activity frequency, from −0.171 for low frequency to −0.439 for high frequency. This indicates a dose-response relationship. Menhas et al. [8] provided specific path coefficients, such as 0.631 (between physical activity and psychological resilience) and 0.601 (between physical activity and quality of life), which quantified the strength of these relationships. Abouzeid et al. [6] found a difference in SpO_2 of 2% between the experimental and control group.

2.8 Consistency of findings

Liu et al. [9] found that males meeting PA guidelines had significantly lower odds of depression symptoms (OR = 0.73), showing an inverse relationship between physical activity and depression, but this association was not significant for females. However, the data in the study by Liu et al. [9] also demonstrated that both males and females who engaged in regular physical education had a decrease in both feelings of loneliness and insomnia. Moreover, the authors of Saqib et al.'s [11] study refer to studies that show that increased physical activity was associated with improved mental health.

2.9 Comparison of effect sizes

While effect sizes like Cohen's d were not directly calculated, the regression coefficients from Liu et al. [9], the path coefficients from Menhas et al. [8], and the odds ratios from Liu et al. [9], along with the physiological data from Abouzeid et al. [6] provide comparable metrics for analyzing the strength and direction of different interventions and the relationship between PA and MH outcomes across studies.

2.10 Moderating factors

Liu et al. [9] indicated that the association between recreational PA and mental health was significant for males but not females, highlighting the potential moderating role of sex. This was further investigated by examining the results of a study they cite, which noted that males and females who participated in regular physical education experienced a decrease in feelings of loneliness and insomnia. Menhas et al. [8] explored moderating factors such as social support, but such factors were not further investigated in their study.

2.11 Statistical analysis

Descriptive and inferential statistics were used to analyze the relationship between physical activity and mental health, drawing from the quantitative findings of the four key studies. These analyses included:

2.11.1 Regression analysis

Linear regression coefficients from Liu et al. [9] were compared to assess how different frequencies of PA were related to mental health outcomes.

2.11.2 Path analysis

The path coefficients reported by Menhas et al. [8] were used to understand the direct and indirect effects of PA on mental health *via* psychological resilience and quality of life.

2.11.3 Odds ratio analysis

The OR reported by Liu et al. [9] (0.73 for males) was used to quantify how PA affects the likelihood of experiencing depression symptoms.

2.11.4 Comparative analysis

Physiological data from Abouzeid et al. [6] (e.g., SpO_2 levels) were analyzed to compare the impact of different training methods on physiological outcomes.

3. Mental health of adolescents

Consistent physical activity is crucial for both physical and mental health during adolescence [11]. Active teens show improved mental well-being, with a positive link between physical activity and resilience [9]. The World Health Organization recommends at least 60 minutes of moderate-to-vigorous exercise daily for children and adolescents [11]. However, around 80% of teenagers aged 13–15 globally fail to meet these activity levels [8], and over 77% of adolescents in China do not achieve the recommended exercise [8]. The rise of motorized transport and sedentary lifestyles contributes to this lack of activity, making it essential to promote active habits among youth.

Consequently, addressing mental health during adolescence is crucial, as this developmental period is when many mental health disorders first emerge, often persisting into adulthood and leading to long-term challenges [12]. The study by Ormel et al. [12] highlights that over half of lifetime mental disorders occur during adolescence, with a significant proportion experiencing multiple co-occurring disorders, which often exacerbates severity. Lifetime mental disorders encompass various mental health conditions diagnosed at any stage of life. These conditions include mood disorders (such as major depression, dysthymia, and bipolar disorder), anxiety disorders (like panic disorder, social phobia, and generalized anxiety disorder), and substance use disorders (including alcohol and drug abuse or dependence). The diagnoses can be made using structured diagnostic tools like the Composite International Diagnostic Interview 3.0 (CIDI 3.0) [13]. Early interventions are essential, particularly given that temperamental traits like neuroticism and poor self-regulation contribute to the development of mental health issues. Physical activity has been shown to play a key role in improving adolescent mental health by reducing symptoms of anxiety and depression, enhancing self-esteem, improving emotional regulation, and providing stress relief. Additionally, it helps prevent or mitigate the onset of co-morbidity, which is common in adolescents with mental health issues, and can improve overall resilience in the face of stress. By fostering social connections, physical activity also serves as a preventive tool, offering a holistic approach to managing mental health that may help prevent future chronicity and reduce the long-term impact of mental disorders. Given the evidence for both the high prevalence of mental health challenges and the benefits of physical activity, integrating exercise into daily routines for adolescents could be an effective strategy for early intervention, reducing both the severity and persistence of mental health issues into adulthood.

4. Physical activity and mental health of adolescents

The mental health of adolescents has become an increasingly important area of focus in public health, as this developmental stage is marked by heightened vulnerability to stress, anxiety, and mood disorders. In recent years, physical activity has been recognized not only for its physical benefits but also for its potential positive effects on mental health.

Among the different forms of exercise, high-intensity interval training (HIIT) has gained attention for its ability to improve cardiorespiratory fitness, enhance pulmonary function, and boost overall physical performance. The study conducted by Abouzeid et al. [6] investigated the impact of an 8-week HIIT program with and without the use of elevation training masks (ETMs) on the physical health markers of young, active adolescents. While the study primarily focused on physiological outcomes such as improved pulmonary function, hematological markers, and cardiorespiratory fitness, these improvements likely contribute to the overall mental well-being of adolescents.

Improvements in physical fitness, including increased VO_2 max (higher VO_2 max means your heart and lungs are more effectively supplying blood to the muscles and are efficiently extracting and using oxygen from the blood), peak power output (PPO: the highest level of power that can be generated under normal conditions before any restrictions or limitations are imposed), and enhanced pulmonary function, have been consistently linked to better emotional regulation and cognitive function [14–16]. In this study, both the experimental group (using ETMs) and the control group showed significant improvements in their fitness levels, with the experimental group experiencing greater enhancements.

Although Abouzeid et al. [6] study did not directly measure mental health outcomes, the physiological benefits achieved through HIIT—such as increased oxygen uptake, enhanced cardiovascular performance, and improved hematological variables—are known to promote better mood, reduce stress, and potentially alleviate symptoms of depression and anxiety.

The physiological mechanisms behind these improvements suggest that regular physical activity, particularly intense, structured exercise like HIIT, can activate positive hormonal responses, including endorphins and other neurochemical factors that contribute to emotional well-being [7]. Moreover, the findings of Abouzeid et al. [6] support the hypothesis that physical activity enhances self-esteem and self-confidence in adolescents, as they observe tangible improvements in their physical performance. These factors, when combined, may lead to enhanced psychological resilience, reduced perceived stress, and overall improved mental health in young people.

As physical health is inextricably linked to mental health, the findings of this study offer valuable insights into the broader benefits of exercise for adolescents. The enhanced physical fitness achieved through HIIT could be a useful tool in addressing the mental health challenges faced by today's youth, promoting not only physical but also psychological well-being.

5. The role of physical activity type and sports participation in mental health

A significant body of evidence supports the positive association between sports participation—particularly team sports—and improved mental health outcomes for adolescents. Guddal et al. [17] found that participation in team sports was linked to higher self-esteem, life satisfaction, and lower psychological distress, particularly as adolescents transition to senior high school. Similarly, McMahon et al. [18] demonstrated that team sports are associated with lower levels of depression and anxiety, with the most significant benefits for girls, particularly in enhancing their mental health compared to individual sports.

Zulyniak et al. [19] extended these findings by suggesting that recreational physical activity, such as organized sports, was positively correlated with better mental health, particularly in male adolescents. In contrast, non-recreational physical activity (e.g., chores) was found to have minimal benefits on mental health, further emphasizing the importance of engaging in structured, social forms of exercise for youth. Similarly, Triaca et al. [20] examined the role of physical education in high school and found that participation in physical activity through school programs significantly reduced insomnia and loneliness, underscoring the mental health benefits of team-based school sports.

Murphy et al. [21] highlighted the critical role of physical activity in improving mental health, particularly for those adolescents who engage in team sports. The study noted that participation in sports was associated with lower levels of depression and anxiety, with team sports showing the most pronounced benefits. This suggests that the social and structured nature of team sports plays a crucial role in enhancing mental health, especially for those who are least active.

Barone et al. [22] demonstrated that even short-term initiatives like the European Week of Sport (EWoS) encouraged previously inactive children to engage in physical activity, which led to sustained improvements in mental health. These findings suggest that both the structure and social aspects of team sports are vital for supporting adolescent mental health and that physical activity interventions should focus on fostering social connections and enjoyment.

6. Gender differences in the relationship between physical activity and mental health

The relationship between physical activity and mental health shows notable gender differences. Gómez-Baya et al. [23] highlighted that girls reported higher levels of depressive symptoms and were less active than boys, with the effect of physical activity on mental health being more pronounced for boys. Shennar-Golan and Walter [24] also pointed out that boys were more likely to engage in strenuous physical activity and that this engagement was influenced by the parent-adolescent relationship, whereas girls tended to be motivated by health and appearance concerns.

Murphy, McGrane, et al. [25] and Zulyniak et al. [19] both found that, while physical activity improved mental health in both genders, the benefits were often stronger in males, particularly with recreational and team-based activities. Conversely, girls did not demonstrate significant associations between recreational physical activity and mental health, although this may be due to limitations in statistical power. These findings suggest that interventions should account for gender differences, tailoring strategies to engage girls through health-related motivations and boys through structured, competitive sports.

Harbec et al. [26] also found gendered differences in the trajectory of emotional distress in childhood, with boys benefiting more from early engagement in sports to avoid increasing emotional distress. This emphasizes the need to consider sex-specific interventions and the unique factors influencing activity levels and mental health outcomes in boys and girls.

7. The frequency and duration of physical activity: Impact on mental health

The frequency and duration of physical activity (PA) have significant effects on adolescent mental health, with several studies demonstrating that more frequent and sustained engagement in PA correlates with improved emotional well-being.

Gómez-Baya et al. [23] conducted a longitudinal study that showed that increased frequency of sport participation was consistently associated with lower levels of depressive symptoms over time. The study highlights a progressive increase in depressive symptoms during adolescence, with less active adolescents reporting higher depressive symptoms. This underscores the need for promoting consistent physical activity to counteract emotional distress.

In a similar vein, Guddal et al. [17] found that adolescents who engaged in regular physical activity, particularly through team sports, exhibited higher levels of self-esteem and life satisfaction. Their study also revealed that PA had a stronger effect on reducing psychological distress among senior high school students, highlighting the importance of maintaining consistent PA during the transition from junior to senior high school.

Murphy, Sweeney, et al. [27] also explored the optimal frequency of PA and its influence on mental health. Their study on a lunchtime games-based intervention found that attending physical activity sessions twice weekly led to significant reductions in anxiety and depression, while attending three times weekly did not provide additional benefits. This suggests that a moderate frequency of activity, rather than excessive engagement, may be most effective for promoting mental health outcomes, particularly for adolescents dealing with anxiety and depressive symptoms.

Murphy, McGrane, et al. [25] further emphasize the role of PA in improving adolescent mental health by highlighting the psychological benefits of PA participation, with a specific focus on intrinsic motivation and social support. While the study primarily explores the social aspects of PA, it also underscores that regular engagement in physical activity—particularly in group settings—provides sustained mental health benefits. Their findings suggest that adolescents who engage in frequent and enjoyable physical activity experience improvements in well-being, resilience, and emotional stability, aligning with previous research that emphasizes the importance of duration and regularity in PA.

The findings from Barone et al. [22] further support the positive influence of PA, particularly through short-term, high-intensity events such as the European Week of Sport (EWoS). The study demonstrated that even brief exposure to increased PA levels can motivate adolescents to continue engaging in physical activity, thereby promoting long-term mental health benefits. Harbec et al. [26] also found that early participation in sports was associated with reduced emotional distress trajectories, suggesting that sustained PA, especially during early childhood, has lasting benefits on emotional health. Taken together, these studies illustrate that physical activity, especially when frequent and of moderate duration, has a measurable impact on adolescent mental health. Whether through structured school-based interventions [27], organized sports [17], or community engagement [22], fostering regular participation in PA can play a pivotal role in mitigating mental health issues such as anxiety, depression, and stress. The integration of both moderate intensity and duration, as highlighted in several studies, appears to be crucial in maximizing the mental health benefits of PA.

8. The role of family and social support in promoting physical activity and mental health

The importance of social support and family involvement in promoting physical activity (PA) and enhancing mental health during adolescence is well-documented across multiple studies. Murphy, McGrane, et al. [25] conducted a study focusing on how adolescents perceive the relationship between PA and well-being, with a particular emphasis on social factors like social interaction and group settings. Their findings suggest that participation in team sports and group activities can foster a sense of belonging, which is crucial for maintaining motivation and enjoying physical activities. This sense of community and social connection significantly contributes to mental health and emotional stability. The study further found that PA offers a pathway to mindfulness and flow states, particularly in outdoor activities, which can significantly benefit adolescent well-being. Murphy, McGrane, et al. [25] highlight that social support, especially through intrinsic motivation and peer encouragement, plays a pivotal role in adolescents' physical activity participation and, in turn, their mental health outcomes.

Zeqiri et al. [28] emphasized the impact of social support from family, friends, and teachers on increasing physical activity levels among adolescents in Kosovo. Their study found that positive encouragement from parents, particularly in organizing activities and providing transport to sports facilities, played a significant role in fostering an active lifestyle. The study suggests that family support is a key factor in overcoming barriers to PA and ensuring adolescents remain active, which aligns with the findings of Murphy, McGrane, et al. [25] on the importance of social support in maintaining regular PA participation.

Shennar-Golan and Walter [24] also explored how family dynamics and relationships influence adolescents' engagement in strenuous physical activity, finding that the parent-adolescent relationship was significantly associated with participation in vigorous PA. This study supports the idea that a positive family environment is crucial for encouraging physical activity, which is linked to improved mental health outcomes, particularly in boys who may be more motivated by familial relationships.

Triaca et al. [20] further add to this body of evidence by examining the role of physical education in the mental health of Brazilian adolescents. Their findings show that adolescents who participated in physical education reported lower rates of loneliness and insomnia, with the effect being more pronounced in boys. The study underlines the importance of structured family and school support systems in promoting physical activity, suggesting that maintaining physical education as a core subject in the school curriculum is essential for supporting adolescent mental health.

Murphy, Coulter, et al. [29] contribute to this narrative by focusing on how self-determined motivation, rooted in family and social contexts, can enhance the mental health benefits of physical activity. The study highlights that intrinsic motivation, driven by autonomy, competence, and relatedness (all core elements of self-determination theory), is crucial in fostering positive mental health outcomes through physical activity.

9. Cognitive and psychological benefits of physical activity

Physical activity is not only beneficial for mental health but also plays a crucial role in cognitive development. Amornsriwatanakul et al. [30] found that regular physical activity enhanced cognitive functions such as memory, attention, and executive function. This aligns with Murphy, Coulter, et al. [29], who found that physical activity in the context of self-determination theory (SDT) led to better psychological outcomes, including enhanced feelings of competence and self-esteem, which are critical for positive mental health. These findings highlight the importance of incorporating cognitive stimulation and psychological satisfaction into physical activity programs for adolescents, further supporting the need for structured and enjoyable activities that foster cognitive and emotional growth.

10. Quality of life

Adolescence is a critical developmental stage where PA plays a crucial role in shaping both mental health and overall quality of life (QLAH). Research consistently shows a strong connection between physical activity and improved quality of life, particularly through its effects on mental health, stress, and emotional stability.

The link between physical activity and mental health is especially relevant in the context of anxiety, distress, and peer relationships. A regression analysis conducted in China found that high-frequency physical activity is significantly associated with lower anxiety levels, while even medium-frequency activities have been shown to reduce distress in adolescents. These findings suggest that engaging in regular PA has the potential to positively influence psychological well-being by mitigating symptoms of anxiety and obsessive-compulsive behaviors, thus promoting emotional stability and a healthier personality [31].

Moreover, the importance of social support in maintaining adolescent mental health cannot be overstated. As demonstrated in research on Spanish adolescents, when family members and friends engage in physical activity together, it not only improves personal PA levels but also enhances family life satisfaction and the overall quality of life. This interconnection between social relationships, physical activity, and well-being underscores the multifaceted nature of quality of life, wherein supportive family and peer environments contribute significantly to adolescents' physical and mental health [32]. Adolescents who have strong, positive relationships with their families and friends report higher levels of life satisfaction and are more likely to engage in regular PA, which in turn reinforces their overall health outcomes.

In the context of the COVID-19 pandemic, the disruptions to daily routines—including the abrupt closure of schools and sports—illustrated how significant physical activity and social interaction are for adolescent mental health and quality of life. Adolescents experienced increased anxiety, fatigue, and poorer sleep quality during lockdowns, emphasizing the negative impact of decreased physical activity and social isolation on mental and emotional health. Interestingly, the quality of sleep emerged as a stronger predictor of improved mental health outcomes, such as reduced fatigue and better peer relationships, than physical activity itself. This suggests that while physical activity is crucial, other factors, particularly sleep quality, can also play a protective role in preserving adolescent well-being during challenging times [33].

Taken together, these studies illustrate the intricate relationship between physical activity, mental health, and quality of life in adolescents. Regular PA—especially when supported by strong familial and social ties—can enhance emotional well-being, reduce symptoms of anxiety and distress, and improve sleep quality, which in turn supports better peer relationships and overall life satisfaction. However, as highlighted during the pandemic, disruptions to these routines can have significant adverse effects, underscoring the need for strategies that prioritize both physical activity and mental health support in challenging times.

11. Aspects of normal development

Understanding the normal development of adolescents in all aspects—cognitive, emotional, social, and autonomous—is crucial when using physical activity as a preventive approach for mental health issues. During adolescence, individuals undergo significant changes that influence their behavior and well-being. Cognitive development allows them to think abstractly and reflect on their identity, while emotional growth leads to mood swings and sensitivity to social pressures [34]. Socially, adolescents begin to prioritize peer relationships, which can impact their sense of belonging and self-esteem. The desire for autonomy often leads to risk-taking behaviors, but also to opportunities for developing healthy habits, such as regular physical activity. By understanding these developmental stages, clinicians and parents can better support adolescents in navigating these challenges and harnessing physical activity as a tool for preventing mental health issues like anxiety, depression, and stress. Knowledge of normative adolescent development is essential to guide appropriate interventions and promote positive behaviors that can safeguard mental health throughout this crucial stage.

In addition to the cognitive, emotional, social, and autonomous developments traditionally considered, we propose a modification to Bagley [34] framework by adding physical development, informed by established normative data on physical potential. This additional perspective allows for a more comprehensive understanding of how physical activity interacts with other developmental aspects to shape overall well-being. A figure illustrating all these aspects of adolescent development—cognitive, emotional, social, autonomous, and physical—has been provided in detail in **Figure 1**. By integrating this new layer of physical development, we aim to enhance the potential for using physical activity not only as a tool for mental health prevention but also as a vital part of adolescent development as a whole.

12. Physical normative data

Understanding the normative data of physical potential is essential for evaluating and interpreting the physical abilities of individuals within a specific population. By relying on robust, well-known, and established measures, researchers and practitioners can obtain reliable benchmarks that reflect typical physical performance across various age groups, genders, and cultural backgrounds. These data are critical for identifying deviations from the norm, assessing health and fitness levels, and developing evidence-based interventions or programs aimed at improving physical health. Accurate normative data also enable professionals to track progress, predict future outcomes, and ensure that individuals receive the appropriate support for their physical development, making it a foundational element of health and fitness assessments. In this context, there are two different types of assessments; health-related physical fitness tests (**Table 1**) and sport-related physical tests (**Table 2**).

13. Proposing a new model utilizing physical activity as preventive medicine for adolescent psychosocial health and well-being

Adolescence represents one of the most critical periods of human development, marked by rapid and often turbulent changes across cognitive, emotional, social, and

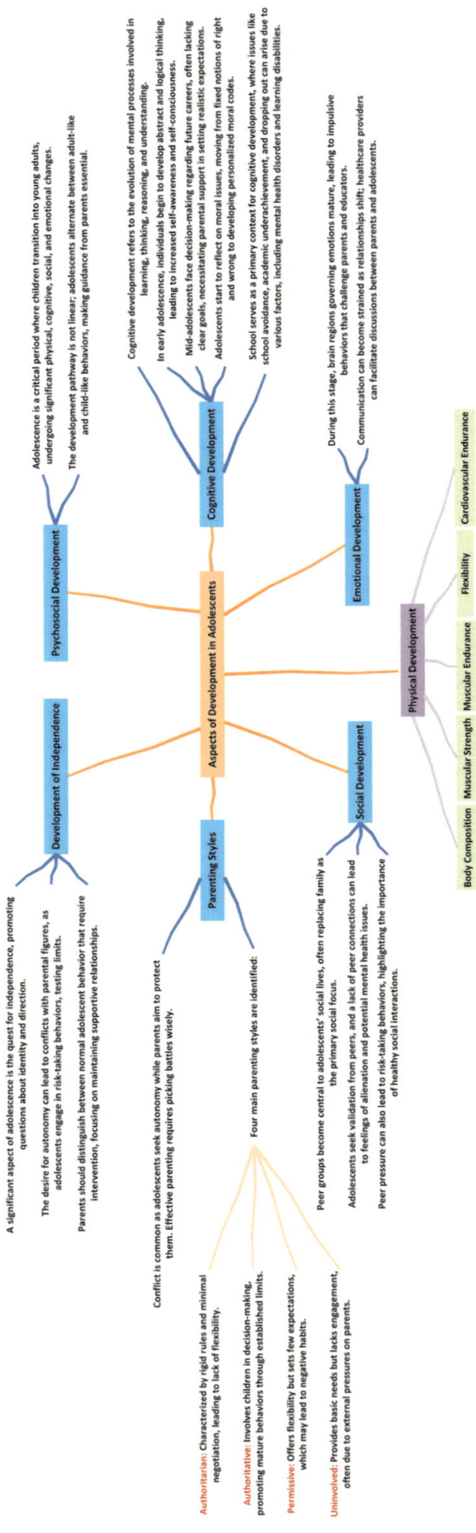

Figure 1.
Aspects of development in adolescents.

Body composition	Muscular strength	Muscular endurance	Flexibility	Cardiovascular endurance
BMI	Dynamometer	Push-up test	Sit and reach test	Step test
Skin fold		Sit-up 60 seconds test	Shoulder mobility	Beep test
Seven site skinfold measurements			BACK test	Yo-Yo test
Three site skinfold measurements			Deep squat	

Table 1.
Health-related physical tests.

Power	The agility	The speed	Speed endurance	Coordination	Accuracy	Balance
Medicine ball throw test	Illinois agility test	30 m dash test	Margaria test	Ball throwing and receiving test	Pass test on overlapping circuits	Y balance test
Sargent test	The pro agility (5-10-5) test		Intermittent fitness test	The numbered circles test illustrates the eye-foot coordination 2.4.2. Eye-hand coordination test		
	Hexagon agility test		30–15 intermittent fitness test (30–15 IFT)			
	Agility T test		Wingate anaerobic test			

Table 2.
Sport-related physical tests.

physical domains. It is a time when young individuals move from dependence to autonomy, navigating complex transitions that significantly impact their mental health and psychosocial well-being. However, the typical psychosocial challenges faced during this period—such as identity crises, peer pressure, academic stress, and emotional instability—often leave adolescents vulnerable to mental health disorders, including anxiety, depression, and eating disorders. While traditional interventions such as psychotherapy and pharmacotherapy remain vital, emerging research points to the promising role of physical activity and sports as effective, evidence-based preventive measures against these challenges. Here, we propose a model that outlines a new framework for leveraging physical activity and sports as preventive medicine for adolescent psychosocial health, drawing upon established developmental principles and contemporary findings.

13.1 The Integrated Developmental Model (IDM) of physical activity as preventive medicine

At the core of the IDM lies the understanding that adolescent development occurs holistically, with physical, cognitive, emotional, social, and autonomous aspects deeply intertwined. By focusing on these interconnected dimensions, physical activity—when thoughtfully integrated into adolescents' daily lives—becomes a powerful tool not just for improving physical fitness but also for enhancing psychosocial health. Physical activity can be a strategic intervention, targeting multiple developmental challenges adolescents face, simultaneously improving physical health and psychosocial well-being.

13.1.1 Cognitive development and physical activity

Cognitive maturation during adolescence includes an increase in abstract thinking, self-reflection, and a deeper understanding of one's values and beliefs [34]. However, adolescents often struggle with maintaining focus and motivation, leading to academic challenges and self-doubt. Integrating physical activity into daily routines can help promote cognitive development in several key ways. Exercise has been shown to stimulate neuroplasticity, supporting memory, attention, and executive function—all critical components of adolescent learning. Furthermore, physical activity helps alleviate cognitive overload, which often accompanies the stress of academic performance and social expectations. Adolescents who engage in regular physical activity are better equipped to manage stress and maintain focus, which can prevent cognitive-related mental health issues, such as anxiety and academic burnout.

13.1.1.1 Integrative cognitive-sport programs

Schools and communities should incorporate "brain breaks" or physical activity sessions within the academic environment. For example, incorporating structured physical activity sessions (such as 15-minute aerobic exercises or sports) in the middle of academic lessons can enhance students' cognitive function, reduce anxiety related to learning, and improve academic performance. These sessions should aim to promote physical coordination while also engaging the mind with mindfulness techniques or problem-solving activities.

13.1.2 Emotional development and physical activity

Adolescence is characterized by significant emotional fluctuations driven by the maturation of brain regions responsible for emotional regulation [34]. This emotional lability often manifests as mood swings, irritability, and heightened sensitivity to social stressors, all of which contribute to the vulnerability of adolescents to mental health disorders, particularly depression and anxiety. Physical activity—especially team sports—can play a crucial role in stabilizing emotions and enhancing emotional resilience. Exercise is known to trigger the release of endorphins, which are natural mood boosters, and it also helps adolescents develop healthier coping mechanisms in response to stress. Moreover, engaging in physical activity fosters emotional self-regulation, a skill necessary for navigating the intensity of adolescence.

13.1.2.1 Emotionally regulated sports programs

We recommend the design of physical activity programs that include a focus on emotional self-regulation, especially in group settings. Programs like "Mindful Movement Sports" can combine aerobic exercises with mindfulness practices to help adolescents regulate their emotions, enhance self-awareness, and reduce emotional reactivity. For example, yoga combined with team sports can teach adolescents both physical coordination and emotional grounding, making them more resilient to psychosocial stress.

13.1.3 Social development and physical activity

During adolescence, peer relationships come to the forefront, with adolescents prioritizing social validation and identity formation through interactions with their peers [34]. This shift can lead to increased susceptibility to peer pressure, bullying, and social exclusion, which often contribute to depression and anxiety. Physical activity, particularly in the form of team sports or community-based exercise groups, creates a structured environment where adolescents can develop positive social skills, enhance self-esteem, and form supportive friendships. Sports provide a platform for adolescents to experience belonging, cooperation, and acceptance, fostering a sense of community that can buffer against the risks of social isolation and loneliness.

13.1.3.1 Social-integration sports teams

Schools and community centers should implement inclusive, non-competitive team sports or fitness programs that encourage participation from adolescents of all skill levels and backgrounds. These programs should focus on cooperation, mutual respect, and social inclusion rather than competition. An example would be mixed-gender sports teams or inclusive fitness challenges that emphasize teamwork and camaraderie, encouraging adolescents to engage with peers from diverse social circles.

13.1.4 Autonomous development and physical activity

Autonomy is a central developmental task during adolescence, as young people begin to form their own identities and make independent choices [34]. However, this drive for independence often leads to experimentation with risk-taking behaviors, such as drug and alcohol use, unsafe sexual practices, and reckless driving. Physical activity offers a constructive outlet for this desire for autonomy by providing adolescents with opportunities to make independent choices about their health, fitness, and well-being. Through sports, adolescents can gain a sense of control over their bodies, learn responsibility, and develop the self-discipline needed to make healthy lifestyle choices.

13.1.4.1 Autonomy-enhanced physical programs

Programs aimed at promoting autonomous decision-making in sports should be emphasized. One example is creating "self-directed fitness plans" where adolescents can set their own fitness goals and track progress with minimal external oversight. These programs can be integrated into schools, community sports clubs, or even after-school fitness initiatives. Adolescents could be encouraged to take leadership roles in coaching, organizing events, or helping younger participants, which would build their confidence and reinforce responsible behavior.

13.1.5 Physical development and psychosocial health

The integration of physical development, guided by normative data on physical potential, enhances the IDM by incorporating the biological aspect of adolescence.

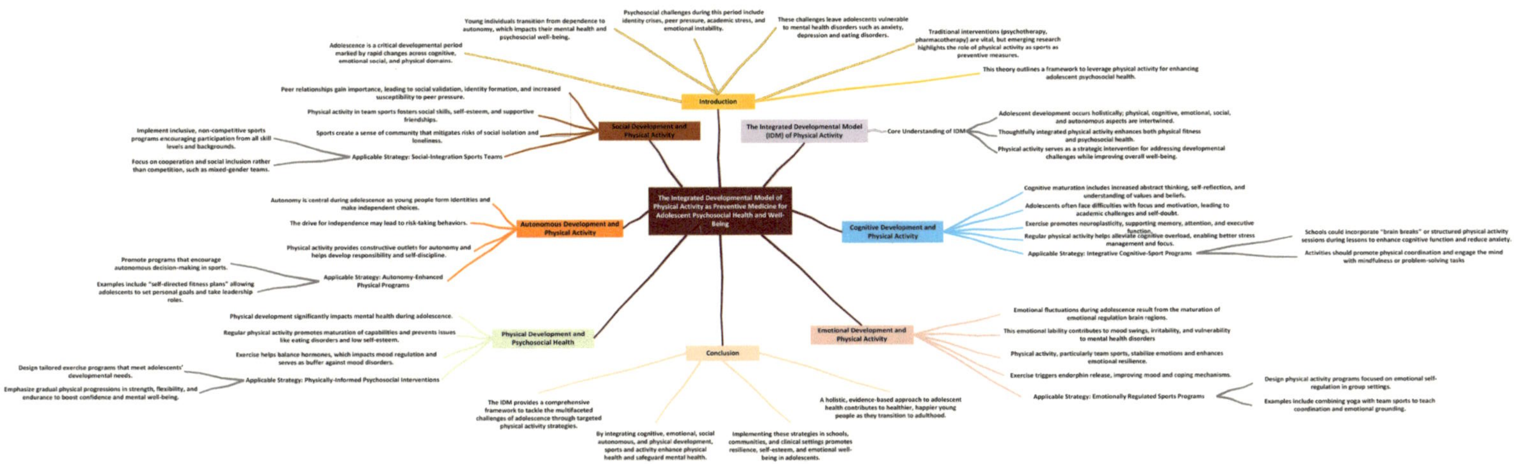

Figure 2.
The Integrated Developmental Model (IDM) of physical activity as preventive medicine for adolescent psychosocial health and well-being.

Adolescents undergo significant changes in strength, endurance, and body composition, which directly impact their mental health. Engaging in physical activity during this period of development is critical not only for the maturation of physical capabilities but also for ensuring a positive body image and preventing issues like eating disorders, body dysmorphia, and low self-esteem. Furthermore, regular exercise can help balance hormone levels, which have a direct impact on mood regulation, offering a physiological buffer against mood disorders such as depression and anxiety.

13.1.5.1 Physically informed psychosocial interventions

Tailored exercise programs, based on established data on physical potential, should be designed to meet the developmental needs of adolescents. These programs should emphasize gradual physical progressions in strength, flexibility, and endurance, ensuring that adolescents feel confident in their growing capabilities. This could include supervised resistance training, cardiovascular endurance programs, and flexibility routines, which are known to promote both physical development and mental well-being.

The Integrated Developmental Model (IDM) of Physical Activity as Preventive Medicine for Adolescent Psychosocial Health and Well-Being offers a comprehensive framework for addressing the multifaceted challenges of adolescence through targeted physical activity strategies. By incorporating cognitive, emotional, social, autonomous, and physical development into a unified approach, we can use sports and physical activity not only to enhance physical health but also to safeguard against mental health issues, fostering resilience, positive self-esteem, and emotional well-being in adolescents. These strategies, when implemented in schools, communities, and clinical settings, provide a holistic and evidence-based approach to adolescent health, paving the way for healthier, happier young people as they transition into adulthood.

The Integrated Developmental Model (IDM) of Physical Activity as Preventive Medicine for Adolescent Psychosocial Health and Well-Being highlights the interconnectedness of cognitive, emotional, social, autonomous, and physical development during adolescence. It emphasizes how physical activity can address these developmental aspects simultaneously, promoting mental health and resilience. The model integrates evidence-based strategies that use physical activity and sports to enhance cognitive function, emotional regulation, social skills, autonomy, and physical health, while preventing mental health issues such as anxiety, depression, and low self-esteem. A detailed illustration of this model is provided in **Figure 2**.

14. Conclusion and recommendations for future research

14.1 Conclusion

This book chapter reveals that physical activity plays a vital role in promoting adolescent mental health, with significant variations based on the type, frequency, and context of the activity. Team sports and recreational physical activities emerge as particularly beneficial, especially when social support and intrinsic motivation are emphasized. Gender differences are critical considerations, with boys generally experiencing stronger benefits from physical activity in terms of mental health and physical fitness. Additionally, sleep quality, family support, and cognitive development are closely linked to physical activity, suggesting that a holistic approach—considering psychological, social, and biological factors—is necessary for effective mental health interventions in adolescence.

14.2 Future research

For the recommended future research, an important aspect would be to explore causal pathways and long-term effects of physical activity on adolescent mental health, with a particular focus on developing gender-sensitive, family-oriented, and context-specific interventions to maximize benefits across diverse populations. With respect to the type of studies, the following is recommended:

14.2.1 Longitudinal studies

Essential to understand long-term PA effects on mental health and well-being.
Needed to determine causal relationships over time.

14.2.2 Intervention studies

Necessary to show cause-and-effect by manipulating PA.
Should focus on effectiveness of PA as treatment and optimizing programs.

14.2.3 Diverse populations

Crucial for generalizing findings across age, culture, and socioeconomic groups. The impact of PA across varied settings should be investigated.

14.2.4 Objective measures

Important for validating self-reported data with tools like accelerometers. Both objective and qualitative data should be used for a richer understanding.

14.2.5 Moderating factors

It is necessary to explore how social, environmental, and genetic factors affect PA outcomes. We should examine factors like social support.

14.2.6 Other recommendations

14.2.6.1 Qualitative studies

To understand individual experiences.

14.2.6.2 Neurobiological studies

To explore brain mechanisms.

14.2.6.3 Studies on different PA types

To know their psychological effects.

14.2.6.4 Real-world studies

For effective program implementation.

Acknowledgements

The authors would like to thank the Deanship of Scientific Research at King Faisal University in Saudi Arabia for sponsoring this book chapter with Grant Number (KFU250684). The authors acknowledge the use of AI including ChatGPT for language polishing of the manuscript.

Conflict of interest

The authors declare no conflict of interest.

Author details

Nasser Abouzeid[1*] and Abdullah Aldughaysh[2]

1 Department of Physical Education, College of Education, King Faisal University, Al-Ahsa, Saudi Arabia

2 Department of Special Education, College of Education, King Faisal University, Al-Ahsa, Saudi Arabia

*Address all correspondence to: nibrahim@kfu.edu.sa

References

[1] Finley MI, Pleket HW. The Olympic Games: The First Thousand Years. Mineola, NY: Dover Publications; 2005

[2] Kennell NM. The Gymnasium of Virtue: Education & Culture in Ancient Sparta. Chapel Hill: University of North Carolina Press; 1995

[3] Kyröläinen H. International military sports council (cism)—Sport science commission. Journal of Science and Medicine in Sport. 2017;**20**:S59. DOI: 10.1016/j.jsams.2017.09.097

[4] Tountas Y. The historical origins of the basic concepts of health promotion and education: The role of ancient Greek philosophy and medicine. Health Promotion International. 2009;**24**(2):185-192. DOI: 10.1093/heapro/dap006

[5] Fiuza-Luces C, Garatachea N, Berger NA, Lucia A. Exercise is the real polypill. Physiology. 2013;**28**(5):330-358. DOI: 10.1152/physiol.00019.2013

[6] Abouzeid N, ELnaggar M, FathAllah H, Amira M. Eight weeks of high-intensity interval training using elevation mask may improve cardiorespiratory fitness, pulmonary functions, and hematological variables in university athletes. International Journal of Environmental Research and Public Health. 2023;**20**(4):3533. DOI: 10.3390/ijerph20043533

[7] Saanijoki T. Neurobiology of Physical Exercise: Perspectives on Psychophysiological Effects and Opioidergic Neurotransmission. Turku: University of Turku; 2018

[8] Menhas R, Dai J, Ashraf MA, Noman SM, Khurshid S, Mahmood S, et al. Physical inactivity, non-communicable diseases and national fitness plan of China for physical activity. Risk Management and Healthcare Policy. 2021;**14**:2319-2331. DOI: 10.2147/RMHP.S258660

[9] Liu R, Menhas R, Saqib ZA. Does physical activity influence health behavior, mental health, and psychological resilience under the moderating role of quality of life? Frontiers in Psychology. 2024;**15**:1349880. DOI: 10.3389/fpsyg.2024.1349880

[10] Kyan A, Tanaka S, Takakura M, Olds T, Schranz N, Tanaka C. Validity of Japanese version of a two-item 60-minute moderate-to-vigorous physical activity screening tool for compliance with WHO physical activity recommendations. The Journal of Physical Fitness and Sports Medicine. 2021;**10**(2):99-107. DOI: 10.7600/jpfsm.10.99

[11] Saqib ZA, Dai J, Menhas R, Mahmood S, Karim M, Sang X, et al. Physical activity is a medicine for non-communicable diseases: A survey study regarding the perception of physical activity impact on health wellbeing. Risk Management and Healthcare Policy. 2020;**13**:2949-2962. DOI: 10.2147/RMHP.S280339

[12] Ormel J, Raven D, Van Oort F, Hartman CA, Reijneveld SA, Veenstra R, et al. Mental health in Dutch adolescents: A TRAILS report on prevalence, severity, age of onset, continuity and co-morbidity of DSM disorders. Psychological Medicine. 2015;**45**(2):345-360. DOI: 10.1017/S0033291714001469

[13] de Beurs D, Ten Have M, Cuijpers P, De Graaf R. The longitudinal association

between lifetime mental disorders and first onset or recurrent suicide ideation. BMC Psychiatry. 2019;**19**(1):345. DOI: 10.1186/s12888-019-2328-8

[14] Wagstaff CRD. Emotion regulation and sport performance. Journal of Sport and Exercise Psychology. 2014;**36**(4):401-412. DOI: 10.1123/jsep.2013-0257

[15] Zhang M, Xv G, Luo C, Meng D, Ji Y. Qigong Yi Jinjing promotes pulmonary function, physical activity, quality of life and emotion regulation self-efficacy in patients with chronic obstructive pulmonary disease: A pilot study. The Journal of Alternative and Complementary Medicine. 2016;**22**(10):810-817. DOI: 10.1089/acm.2015.0224

[16] Zheng F. The impact of long-term endurance training on cognitive function and emotional state in community-dwelling elderly. Journal of Biotech Research (Durant, OK). 2024

[17] Guddal MH, Stensland SØ, Småstuen MC, Johnsen MB, Zwart J-A, Storheim K. Physical activity and sport participation among adolescents: Associations with mental health in different age groups. Results from the Young-HUNT study: A cross-sectional survey. BMJ Open. 2019;**9**(9):e028555. DOI: 10.1136/bmjopen-2018-028555

[18] McMahon EM, Corcoran P, O'Regan G, Keeley H, Cannon M, Carli V, et al. Physical activity in European adolescents and associations with anxiety, depression and well-being. European Child & Adolescent Psychiatry. 2017;**26**(1):111-122. DOI: 10.1007/s00787-016-0875-9

[19] Zulyniak S, Williams JVA, Bulloch AGM, Lukmanji A, Patten SB. Physical activity and mental health: A cross-sectional study of canadian youth. Journal of the Canadian Academy of Child and Adolescent Psychiatry (Guelph, ON). 2020

[20] Triaca LM, Frio GS, França MTA. O Jovem na Educação Básica não Precisa de Educação física? Uma Análise por Gênero do Impacto Dessa Disciplina Sobre a Saúde Mental de Escolares Brasileiros. In: Proceedings of the 46th National Economics Meeting. Rio de Janeiro, Brazil: ANPEC; 2018

[21] Murphy J, Sweeney MR, McGrane B. Physical activity and sports participation in Irish adolescents and associations with anxiety, depression and mental wellbeing. Findings from the physical activity and wellbeing (paws) study. Physical Activity and Health. 2020;**4**(1):107-119. DOI: 10.5334/paah.58

[22] Barone R, Marino Gammazza A, Casarrubea M, De Martino L, Marino Gammazza M, Monachino F, et al. European week of sport: Innovative initiative of European commission that inspires children to be active. The Journal of Sports Medicine and Physical Fitness. 2019;**59**(6):1026-1029. DOI: 10.23736/S0022-4707.18.08458-X

[23] Gómez-Baya D, Calmeiro L, Gaspar T, Marques A, Loureiro N, Peralta M, et al. Longitudinal association between sport participation and depressive symptoms after a two-year follow-up in mid-adolescence. International Journal of Environmental Research and Public Health. 2020;**17**(20):7469. DOI: 10.3390/ijerph17207469

[24] Shennar-Golan V, Walter O. Physical activity intensity among adolescents and association with parent–adolescent relationship and well-being. American Journal of Men's Health. 2018;**12**(5):1530-1540. DOI: 10.1177/1557988318768600

[25] Murphy J, McGrane B, White RL, Sweeney MR. Self-esteem, meaningful experiences and the rocky road—Contexts of physical activity that impact mental health in adolescents. International Journal of Environmental Research and Public Health. 2022b;**19**(23):15846. DOI: 10.3390/ijerph192315846

[26] Harbec M-J, Goldfield G, Barnett TA, Pagani LS. Physical activity as both predictor and outcome of emotional distress trajectories in middle childhood. Journal of Developmental & Behavioral Pediatrics. 2022;**43**(3):159-167. DOI: 10.1097/DBP.0000000000001005

[27] Murphy J, Sweeney MR, McGrane B. The effect of a games-based intervention on wellbeing in adolescent girls. Health Education Journal. 2022c;**81**(4):463-478. DOI: 10.1177/00178969221090583

[28] Zeqiri L, Georgiev G, Gontarev S. Influence of social support and residential area quality on the physical activity of adolescents from the republic of Kosovo. Pedagogical Almanac. 2022;**30**(1):100-104. DOI: 10.54664/USFO1428

[29] Murphy J, Coulter M, Sweeney MR, McGrane B. "You get to..." a qualitative study of perceived influence of physical activity and sport on mental wellbeing among adolescent girls. Advances in Physical Education. 2022a;**12**(02):87-105. DOI: 10.4236/ape.2022.122008

[30] Amornsriwatanakul A, Lester L, Bull FC, Rosenberg M. Are Thai children and youth sufficiently active? Prevalence and correlates of physical activity from a nationally representative cross-sectional study. International Journal of Behavioral Nutrition and Physical Activity. 2017;**14**(1):72. DOI: 10.1186/s12966-017-0529-4

[31] Li W, Gao Y, Liu G, Hao R, Zhang M, Li X. Shifting the paradigm: A fresh look at physical activity frequency and its impact on mental health, life satisfaction, and self-rated health in adolescents. International Journal of Mental Health Promotion. 2024;**26**(2):83-92. DOI: 10.32604/ijmhp.2023.042014

[32] Urchaga JD, Guevara RM, Cabaco AS, Moral-García JE. Life satisfaction, physical activity and quality of life associated with the health of school-age adolescents. Sustainability. 2020;**12**(22):9486. DOI: 10.3390/su12229486

[33] Wingerson MJ, Baugh CM, Provance AJ, Armento A, Walker GA, Howell DR. Changes in quality of life, sleep, and physical activity during COVID-19: A longitudinal study of adolescent athletes. Journal of Athletic Training. 2023;**58**(10):887-894. DOI: 10.4085/1062-6050-0529.22

[34] Bagley S. Psychosocial Development in Adolescents—Pediatrics. Rahway, NJ: MSD Manual Professional Edition; 2024. [cited 2025 Feb 4]. Available from: https://www.msdmanuals.com/professional/pediatrics/health-care-issues-in-adolescents/psychosocial-development-in-adolescents

Section 2

Special Populations and Lifespan Considerations

Chapter 5

Effectiveness of Physical Activity and Sport Interventions on Executive Functioning, Social Skills, and Quality of Life in Children and Adolescents with Neurodevelopmental Disorders

Katerina Asonitou, Sophia Charitou and Dimitra Koutsouki

Abstract

Neurodevelopmental disorders represent a category of psychiatric conditions that manifest during early childhood. They are characterized by atypical development of the central nervous system and can result in a range of emotional and behavioral challenges, as well as significant impairments in psychological, social, academic, and occupational domains. Challenges related to executive function, which include but are not limited to working memory, inhibition, cognitive flexibility, planning and organization, attention, and self-regulation, are commonly observed in individuals with various neurodevelopmental disorders, including autism spectrum disorder (ASD), intellectual disabilities (ID), attention-deficit hyperactivity disorder (ADHD), and developmental coordination disorder (DCD). Furthermore, during childhood and adolescence, executive functioning serves as a crucial predictor of academic success. This chapter examines the impact of diverse exercise and sports intervention programs, adapted appropriately, on individuals with neurodevelopmental disorders. It highlights the benefits of enhancing executive functions and their correlation with improvements in social skills, quality of life, and overall well-being.

Keywords: physical activity, training exercises, intervention, sport, exercise, neurodevelopmental disorders (NDDs), autism spectrum disorder (ASD), intellectual disabilities (ID), attention-deficit hyperactivity disorder (ADHD), developmental coordination disorder (DCD), executive functions, quality of life (QoL), well-being

1. Introduction

"Mens sana in corpore sano" by the Latin poet Decimus Iunius Iuvenalis (55 or 60 Anno Domini-after Christ).

"Νούς υγιής έν σώματι υγιεΐ" in ancient Greek.
"A healthy mind in a healthy body" rendering of the phrase in English.

Neurodevelopmental disorders (NDDs) are inherent conditions rooted in neurological factors, resulting in lifelong challenges such as autism spectrum disorder (ASD), intellectual disability (ID), attention deficit hyperactivity disorder (ADHD), and developmental coordination disorder (DCD), among others. These disorders can lead to difficulties in language, behavior, and learning, often necessitating a multifaceted approach to treatment and intervention [1].

Engaging in consistent physical activity (PA) is essential for sustaining both physical and mental health [2]. Activities that incorporate social interaction, such as active play during school breaks, team sports, or family-oriented physical activities, offer children valuable chances to develop prosocial skills such as effective communication, turn-taking, and conflict resolution. Acquiring these skills can enhance children's enjoyment and involvement in physical activities [3].

The World Health Organization advises that children should engage in at least 60 minutes of moderate-to-vigorous physical activity (MVPA) each day; however, many children worldwide fail to meet this recommendation [4]. Research has shown a positive correlation between children's physical activity levels and their proficiency in fundamental motor skills (FMS). Specifically, children who excel in FMS tend to be more active, while those with lower proficiency are less likely to participate in physical activities. This connection is understandable, as FMS encompass essential skills—such as locomotor, object-control, and stability skills—that are necessary for various forms of physical activity, including sports [5].

Children with complex developmental needs represent a specific group that faces a heightened risk of inactivity and diminished fundamental movement skills [6]. These children may have a single diagnosed developmental disorder, but many experience multiple concurrent developmental challenges [7]. Such conditions can affect their physical capabilities; for instance, DCD is a broad classification that includes motor difficulties stemming from immaturity and developmental delays [8]. Additionally, other conditions can influence their behavioral and social functioning, such as ASD that affects social interactions, communication abilities, repetitive behaviors, and focused interests across a diverse range, [9] as well as ADHD that is marked by inattention, hyperactivity, and impulsive actions [3]. Individuals with intellectual disabilities (IDs) frequently experience low levels of physical fitness, which, combined with reduced physical activity, can result in negative health outcomes. Research consistently shows a link between low physical activity and inadequate cardiovascular fitness, reduced muscular strength and endurance, as well as a higher prevalence of overweight and obesity in both adults and adolescents with ID. Therefore, it is essential to prioritize the creation of strategies that encourage greater participation in physical activities, particularly for children with NDDs, who need to develop lifelong exercise habits [10].

Some children who fail to achieve their developmental milestones may not have an official diagnosis [11]. Addressing these intricate developmental issues is crucial to mitigate the potential for secondary health and psychosocial challenges [12]. Structured and planned physical activity, recognized as exercise, has been shown to be an effective intervention, offering numerous advantages in physical, socio-emotional, behavioral, and cognitive domains (**Figure 1**) [14].

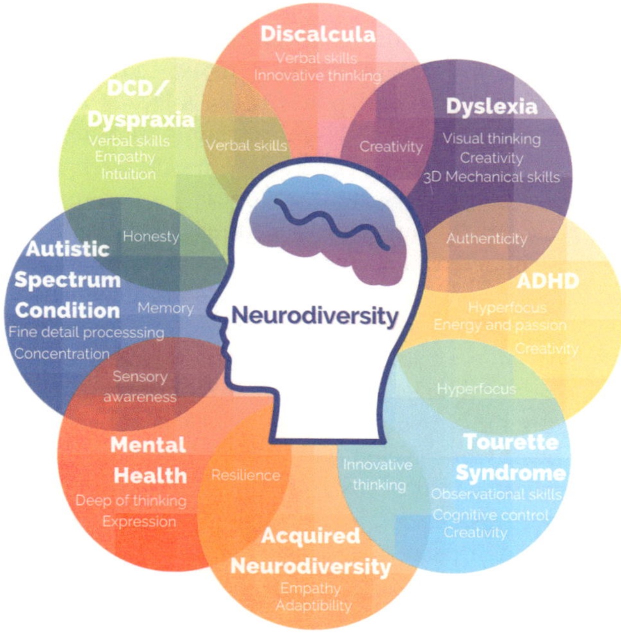

Figure 1.
Understanding neurodevelopmental disorders [13]. https://www.structural-learning.com/post/neurodevelopmental-disorders.

2. Physical activity and exercise for children and adolescents with intellectual disability (ID)

Intellectual disability (ID) is classified as a neurodevelopmental disorder comprised of (1) difficulties in adaptive functioning and (2) problems in intellectual functioning. These limitations can lead to problems functioning in daily life skills relative to people who do not have disabilities [15].

People who have ID generally show a wide decline in cognitive abilities because of organic damage to the brain, or genetic factors, or incomplete development of the brain during early stages of growth. Furthermore, ID is also very heterogeneous, with cognitive deficits differing markedly among people, leading to various profiles of strengths and challenges [16–18]. Intellectual disabilities are not only individuals with at least 70 points below average IQ, but also persisted the individual's as well a family's life. Hence, the areas of rehabilitation, education, and social integration for children with intellectual disabilities become the main focus.

Improving cognitive performance is not only relevant academically; it is essential for personal and psychological growth and is also beneficial for well-being. It is the basis of basic human functions such as reasoning, memory, attention, learning, and problem-solving. Intellectual disabilities frequently involve weaknesses in executive function, attention, memory, mathematical skills, and spatial orientation skills. Consequently, they are often developmentally delayed across multiple domains, including learning, communication, attention regulation, problem-solving, and socialization, all of which contribute to a more complex daily life in comparison with their typically developing counterparts [18]. Moreover, people with disabilities, especially people with intellectual disabilities, are less active than the general population [19]. Low levels of physical activity are common among individuals with intellectual disabilities (ID) and are associated with poor fitness. A lack of fitness can lead to harmful health consequences. Individuals who are adults and adolescents with ID often demonstrate poor cardiovascular fitness, low muscle strength and endurance, and a high prevalence of being overweight and obese. Therefore, the biggest question is how a lack of exercise can affect their health status [10, 16, 19].

Many barriers limit the active participation of people with intellectual disabilities in adapted physical activities. The existing barriers could include poor facilities, insufficient awareness regarding appropriate activities, restrictions due to their disabilities, and unavailability of society support. Another issue includes athletes with intellectual disabilities that experience being underrepresented in community sports and recreation programs limiting opportunities for them to participate in physical activity compared to peers who develop typically. Although the benefits of engaging in physical activity, including improvements in physical fitness and body composition for those with intellectual disabilities, have been extensively studied, the relationships of physical activity with multiple other areas, including behavioral and emotional problems, mental health, and other indicators of psychosocial functioning, remain largely unexplored [19].

Prior research has largely focused on select syndromes (e.g., Down syndrome) or conditions that are commonplace among individuals with intellectual disabilities, such as autism and attention-deficit/hyperactivity disorder [20].

However, these studies often failed to report the level of intellectual disability in their cohorts, which is an important variable since there are strong relationships between the level of participation in physical activity and the severity of intellectual disability [17]. Due to physical disabilities, behavioral issues, or lack of personalized programs that suit their needs, children and adults with profound intellectual disabilities (ID) are often limited in their physical excursion [17, 21].

Exergames integrate innovative technologies to create interactive environments, using gestures and movements of the body instrumental to simulating the gameplay on the screen. The use of a head-mounted display where users travel through fully immersive (FI) virtual reality exergames fully engages users. This technology goes a step further by allowing players to tune out their surroundings and focus solely on the game. Exergames are perceived to be a potential approach to promote physical activity and address typical barriers faced by young individuals with disability. By bringing together video gaming components with physical activity on a single platform, they offer a solid way of promoting movement and fitness [10].

Borland et al. [19] performed a systematic review investigating the potential effects of sports and physical activity on behavioral problems, emotional problems, and psychosocial well-being of children and adolescents with intellectual disabilities. As of February 28, 2021, the researchers exhaustively searched five databases (ERIC, MEDLINE, PsycINFO, SportDISCUS, and SCOPUS). Out of these, a total of 32 studies were included as per the predefined inclusion criteria and analyzed. These studies used different research methods: 15 were case studies, 14 were treatment trials, and 2 were cross-sectional studies and 1 was a cohort study. The results showed that the interventions had a beneficial impact but highlighted that one or more (at least 7 of 14) of the trials had a high risk of bias. Overall, the data strongly suggest that higher levels of physical activity relate positively to beneficial behavioral, emotional, mental, and psychosocial outcomes for people with intellectual disabilities.

In a similar study, Scifo et al. [21] focused on examining the positive effect of Sport Intervention Programs (SIPs) for people with intellectual disabilities. In an effort to shed light on key motivators for SIP participation, they conducted a systematic review of existing literature to provide a comprehensive overview of current knowledge. Through the comparison of 12 distinct studies, they emphasized how SIPs varied according to age and types of disability among participants, as well as the way the interventions were approached and carried out. SIP and identified important enablers for effective engagement in SIP (e.g., enabling environments for SIP (appropriate infrastructures), physical activity programs at schools and communities, and building educational and facilitator capacity). The results obtained demonstrate the importance of integrating successful SIPs to a more general intervention framework designed to improve physical and mental status and social integration of individuals with IDs [21].

Yang et al. [17] built upon the findings of previous research and performed a meta-analysis investigating the impact of physical activity on the mental health of children with intellectual disabilities. This study used the RE-AIM framework (Reach, Effectiveness, Adoption, Implementation, and Maintenance), a critical tool for evaluating the implementation of public health programs. In order to fill the gaps found in the literature, a comprehensive systematic review was performed using six databases (CINAHL, ERIC, PsycINFO, PubMed, SPORTDiscus, and Web of Science). Studies published between January 2000 and September 2021 were reviewed; very strict inclusion criteria were applied, such that each selected study must have included physical activity as an intervention and must have assessed mental health outcomes in children and adolescents aged 5–17 years with intellectual disabilities. This systematic review conducted in accordance with PRISMA guidelines used Comprehensive Meta-Analysis within the RE-AIM framework. This meta-analysis of 15 studies found a statistically significant positive impact of physical activity on the mental health of children and adolescents with intellectual disabilities, with a large effect size. More specifically, the effects showed moderate impact on psychological well-being, as well as significant improvement in cognitive functioning. The positive effects of these results highlight the effect of physical activity on the mental health of children and adolescents with intellectual disabilities, where psychological well-being and cognitive function show improvement. Thus, it is important to evaluate interventions to promote physical activity by using the RE-AIM framework to enable the translation of research evidence into effective practices that assess both short-term and long-term outcomes [17].

Executive dysfunction is common in people with intellectual disabilities (ID). Zhu et al. [18] recently performed a systematic literature review and meta-analysis to systematically investigate the effects of physical activity interventions on cognitive function in children and adolescents with intellectual disabilities. Researchers conducted a systematic search of Pacific and European databases from inception until January 23, 2023, for studies reporting effects of physical activity/exercise interventions on cognitive outcomes in this population. This search resulted in 2393 articles. Following an extensive review and stringent screening procedure, a total of 13 articles consisting of 16 studies were selected for the final analysis. Review Manager 5.4 software was used to assess the risk of bias from the studies, calculate the effects of the interventions, assess heterogeneity between studies, and perform sensitivity analyses. The findings emphasized a globally significant positive link between physical exercise and cognitive ability among children and adolescents with intellectual disabilities. Notably, this extended to commentaries on other cognitive domains, such as memory updating, attentional focus, and attentional allocation. Findings also showed that in comparison with individuals not engaging in physical activity, physical activity did not show statistical significance for specific functions such as attentional switch, attentional span, visual reaction time, and auditory reaction time, despite differences in intervention approach, intervention duration, and intellectual disabilities across studies resulting in heterogeneity [14].

A recent study by Xiao et al. [22] evaluated whether participation in the Special Olympics Unified Sports (UNS) soccer program influences executive function (EF) and prefrontal cortex activation in adolescents with intellectual disabilities (ID). Using fNIRS, the study compared two groups: those who had undergone UNS soccer training (n = 18) and matched controls who did not participate in the program (n = 26). The flanker task, 2-back task, and dots-triangles task were performed to assess the executive function in three cognitive tasks, corresponding to different cognitive constructs.

The results showed that the soccer training group displayed significant superiority in performance on the dots-triangles task and on the 2-back test compared with the control group. Even though the above-mentioned effects were promising, researchers were unsure if R-DLPFC would be more activated during the dots-triangles task in the soccer condition. Notably, the soccer group exhibited significantly reduced R-FPA activation during the 2-back task compared to the control group. This lack of group differences in the flanker task is noteworthy, suggesting that the soccer training program may not influence all dimensions of executive function.

These results indicate that the UNS soccer training program could improve the working memory functions in adolescents with intellectual disabilities, as it is indirectly enhanced by enhanced activation of the R-FPA. However, the effect on R-DLPFC activation is unclear. The findings suggest that participation in UNS soccer training may be effective in delivering improvements in executive function, thereby potentially improving the quality of life of adolescents with ID [22].

In this context, Chueh et al. [23] carried out a systematic review to analyze previous studies that evaluated the effects of a single session of physical activity on cognitive function in individuals with intellectual disabilities. This review provided a thorough analysis of eight separate studies and concluded that moderate-intensity aerobic physical activity positively influenced cognitive domains, including processing speed, inhibition, and working memory.

Despite these findings, the review highlighted a clear gap in the literature with respect to the most effective intensity levels and types of physical activities that translate into benefits for targeted cognitive functions, specifically higher-order executive functions such as planning. This makes it difficult to formulate precise recommendations as there are not available studies. Overall, the results have important implications for parents, educators, and professionals working with teenagers and young adults with mild intellectual disabilities or Down syndrome. Through the design of specific physical activity interventions decided upon intensity and type, caregivers and practitioners can greatly impact daily function and quality of life of this population. This approach emphasizes what may be a potent additive to learning, work performance, and overall behavior management: personalized physical exercise [23].

3. Autistic spectrum disorder (ASD)

Autism spectrum disorder (ASD) encompasses deficits in cognitive, linguistic, communicative, social, and emotional development. They are chronic and severe conditions usually manifesting in early life, often accompanied by cognitive disabilities and co-morbid conditions [15]. Autism spectrum disorder (ASD) is characterized by core impairments that may include the presence of stereotyped and repetitive behaviors, decreased imaginative play, deficits in language and social skills, and restricted behavioral patterns. Although not specifically listed in widely accepted diagnostic manuals, by far the most consistent patterns of motor impairments among individuals with autism include both gross and fine motor skill deficits, as well as positioning/control, so that, for example, postural and balance difficulties, abnormal gait patterns, hypotonia, and motor coordination deficits are widespread among individuals along the spectrum. These motor challenges are known to be a potential risk factor for autism as they often appear in childhood and can last into adulthood.

Integrated approaches are considered to be effective in the mitigation and management of autism symptoms, since this is a complex disorder that is not addressable through single approach. One promising approach is interventions targeting social and communicative skills for several patients, speech therapy, or followed by a larger community for the patients [24–27].

Specialized sports training for autism can help learn specific skills that can be extended into therapeutic settings and improve social and communicative skills. The positive effects of sports on cognitive ability and mental or physical health have been well studied in the general population. In addition, the benefits of sports for an individual with autism spectrum disorder (ASD) are clear, as exercise reduces motor difficulties, increases cognitive skills, and aids emotional and behavioral health. Apparently, engaging in sport is advantageous for cognitive development as well as mental or physical health for both individuals on the spectrum and typically developing individuals [24–26, 28–32].

3.1 Effects of physical activity on executive function in children and adolescents with autism spectrum disorder (ASD)

Extensive literature exists regarding the positive impact of physical activity on executive functions in children with autism spectrum disorders (ASD) [33, 34].

A recent study conducted by Hou et al. [29] undertook a thorough investigation into the influence of different sports on executive function in children with ASD. A comprehensive search in eight online databases, including Web of Science, PubMed, Cochrane, Embase, and CNKI, was conducted to identify randomized controlled trials and quasi-experimental studies that report relevant data. A meta-analysis was conducted within a Bayesian framework, providing a personalized summary of the findings. The review included several major studies that highlighted the relationship between physical activity and executive function. The analysis showed that exercise was associated with a substantial improvement in all three aspects of executive functioning: inhibitory control, cognitive flexibility, and working memory. Cognitive flexibility and inhibitory control showed small-to-medium improvement, though inhibitory control showed greater improvement than cognitive flexibility. But the benefits to working memory, though significant, fell short of a medium effect size. For particular sports, Mini Basketball was effective in improving inhibitory control and cognitive flexibility, while having no effect on working memory. Meanwhile, Ping Pong effectively enhanced cognitive flexibility as well as working memory skills but demonstrated lower efficiency on inhibitory control. The early results also indicated that while Fixed Bicycle had some positive results, its influence on executive function—especially the targeted skills—was markedly weaker, suggesting less optimism in encouraging executive skills in children with ASD. This study also found varying effectiveness for different interventions. Other approaches regarding executive functions were found in activities such as learning bikes, animal-assisted therapy, and exergaming. Cognitive flexibility would probably not result just from these resources. Such approaches as SPARK, Neiyang Gong, and Martial Arts all serve to improve inhibition—one of the strongest predictors of effectiveness of all the interventions tested. The results showed that SPARK and Fixed Bicycle interventions had little effect on working memory, aligning with previous conclusions that different physical engagements target specific areas of executive function [29].

A recent systematic review by Suárez-Manzano et al. [31] screened 21,577 references from numerous databases and concluded 20 longitudinal studies that met inclusion criteria, which were restricted to high-quality interventions. Objective: This review aimed to investigate the impact of physical activity (PA) in patients with autism spectrum disorder (ASD). Interestingly, 64% of these studies had a positive effect of PA on cognition. Similarly, 46% of the studies reported improvements in behavioral and motor skills outcomes, highlighting the positive impact of exercise interventions in this population. Most PA sessions were moderate to vigorous intensity exercises of 15- to 30 -minute duration that had significant positive impacts on a range of domains including cognitive ($r(41) = 0.44$, $p < 0.003$), behavioral ($r(41) = 0.36$, $p < 0.02$), stereotypical ($r(41) = 0.37$, $p < 0.05$), and repetitive behaviors ($r(41) = 0.28$, $p < 0.05$).

Of the 15 studies that specifically examined the long-term effects of physical activity (PA), most focused on aerobic types of activities before or after school time to promote motor skill development. Ten of these studies have reported beneficial effects on all behavior and stereotypical behavior. In contrast, only six studies found clinically meaningful improvements in motor skills suggesting that while PA can encourage good behaviors, its impact on motor skill development is not uniform. Also, none of the studies found any negative associations between PA and outcomes for participants, which is a telling finding. Follow-up analyses did not reveal any significant differences in the effects of PA on the outcomes based on age, suggesting that the effects of PA are similar regardless of developmental stage. Age, however, is an important quantity to consider when designing educational programs so that we can target interventions for the development map of each child. However, studies have demonstrated differences in individuals' responses to exercise in children without autism spectrum disorder (ASD), indicating the importance of age-related factors for the provision of optimal interventions.

Moreover, no substantial differences were observed in PA effects related to gender of the participants. Importantly, most of the studies mainly included male participants as boys are more frequently diagnosed with ASD findings, as girls had lower PA levels than boys, suggesting that PA interventions could potentially have a greater impact on girls, particularly in this lower group (i.e., dose-response effect). This article highlights the need to research and consider other issues that may impact the benefits of physical activity among people with ASD. Factors including body mass index (BMI), time to school commute, and comorbid intellectual giftedness with ASD are likely to have high variability and significance in the outcome of physical activity interventions in this population. Future research exploring the long-term effects of multiple types of physical activity on social skills, communication abilities, sleep quality, and overall well-being would be worthwhile. Further systematic investigations are needed to clarify the short and long effects of physical activity on cognition, behavior, and motor skills of children with ASD. Several studies examined were of limited sample sizes and did not properly control for confounding variables, and caution is warranted in generalizing these results. In the meantime, more robustly designed studies are needed, allowing researchers to draw nuanced insights, ultimately informing the development of successful intervention strategies [31].

Bochun [35] published an article examining the historical context of exergames, as well as their impact on executive functions, energy expenditure, and exercise intensity in young people with autism spectrum disorder (ASD). Repetitive and stereotyped behavior is common in individuals with autism due to difficulties in inhibitory control and cognitive flexibility. Studies show that exergames could thus be used to promote improvement of executive function and these repeating behaviors in ASD children. After only 20 minutes of somatosensory play, participants with autism exhibited a significant reduction in their repetitive behaviors [34].

In another study, Nekar et al. [36] investigated the efficacy of augmented reality game-based cognitive-motor training over the course of 8 weeks in reducing repetitive behaviors of children with ASD and reported significant improvements in repetitive behaviors of the intervention group. In addition, it was found that aimless or stereotypic physical activity was significantly reduced in children with ASD after 12 somatosensory play sessions in the period of 6 weeks. Parallel trends indicate a relationship between physical movements, stereotyped behavior, and potential fatigue from physical exercise, which may also explain this decrease in repetitive stereotyped behaviors.

From the cognitive framework, metacognition is based on executive function, and improvements in metacognitive domains can promote more favorable executive function outputs. After 30 interventions at the Makoto Arena, remarkable increments in metacognitive markers and higher executive profiles were found. Autistic children have exhibited improvements in their ability to initiate, plan, organize, and set up working memory, as well as problem-solving. There were also modest improvements in inhibitory control, cognitive flexibility, and emotional regulation. Children with ASD who underwent a 12-week somatosensory play intervention showed statistically significant improvements in inhibition, working memory, and flexibility scores, with lasting benefits in both inhibition and flexibility over traditional physical education classes.

The executive function levels of ASD children doing exergames were significantly higher than that of children not doing exergames, and the experimental group showed more advancement than the control group. Some studies suggest single player exergames may be better for improving children's executive function, perhaps because of increased social demands in cooperative exergames that can distract children with ASD and impede their executive function performance in wait times. Motor skills have also been associated with executive function because executive function has been connected to brain regions that respond to motor skill development. All the results about the aspects of cognitive function affected ASD participants: The children's executive function increased after having somatosensory one play, which may be due to modifications of cognitive function and physiological structural, and functional neural networks might be related to the physical one. Physical exercises can boost the flow of blood to the brain, thereby promoting the delivery of oxygen and nutrients required for its function. This also helps cognitive abilities, including working memory, task switching, and attention, which in turn improves executive function [35–37].

3.2 Sport and exercise for ASD young's social and motor skills

Many of the children identified with ASD exhibit deficits in social functioning that challenge their interactions with peers on a daily basis. Social skills are vital for a child's growth, especially for those with ASD. Social functioning comprises the specific set of skills that allows children to respond adaptively in various social situations and interact skillfully with others [36].

Children with ASD are known to have social interaction and communication deficits that impact their quality of life [26]. A strategy suggested for addressing these challenges is integrating physical exercise interventions. Hence, the current meta-analysis by Koh et al. [26] has been performed to determine the impact of physical exercise interventions on social skills among ASD children. Eligible studies included randomized controlled trials or quasi-experimental studies that assessed the effect of exercise interventions on the social skills of children with ASD.

Sixteen studies were included in the meta-analysis. The results showed that physical exercise interventions significantly improved social skills in children with ASD. Age was evidenced from the moderator analysis to be an important variable in determining the outcomes of the interventions intended to enhance social skills, as the analysis highlighted those interventions extending beyond 12 weeks led to statistically significant improvement in social skills. Results of this meta-analysis provide compelling evidence for the efficacy of physical exercise interventions to improve the social skills of children with ASD. The moderator analysis highlights the importance of considering both average age and intervention duration when implementing these programs. Such findings stress the role of physical exercise as an attractive tool for improving social skills among them [26].

According to previous results, in a systematic narrative review Levante et al. [25] overviewed studies on the effectiveness of sports training programs that develop and promote social and motor skills in people with ASD.

The systematic review eligibility criteria were developed based on the PICO protocol. Four databases were searched following the PRISMA protocol. A total of 30 articles were reviewed, and their key features were summarized and critically appraised. The findings reveal that there are eight types of sports used in individuals with ASD and most of them are concentrated on two individual sports (swimming and horseback riding). Overall, they found that sports training programs had a positive impact on motor, social skills, and autistic features (i.e., repetitive behaviors) and individual function. The narrative synthesis also highlighted the need for studies that incorporate more girls and preschool-aged children with ASD in the future. Professional experts must be trained in autistic traits and have experience in the specific sport; collaboration with the autistic community to inform the development of training programs for sports professionals is recommended [25].

Many empirical studies have regularly demonstrated that exercise is a well-established and effective intervention that can provide numerous positive effects for children and adolescents who have been diagnosed with autism spectrum disorder (ASD). Here, we present the first systematic review of the impact of exercise across multiple domains in this population, despite this accumulating evidence.

In an effort to fill this gap in the literature, a large systematic review and meta-analysis was recently published by Jia et al. [28] that aimed to systematically synthesize data from exercise interventions targeting children and teenagers with autism. To collect our primary research studies, the researchers started their search in four well-known academic databases. This was followed by filtration of the literature by two independent researchers who screened for use based on our predefined criteria, which stipulated that only studies that specifically analyzed what we were interested in could be included. The systematic review was conducted rigorously according to the guidelines laid out in the Cochrane Handbook for Systematic Reviews of

Interventions, a venerable guide of meeting the needs of evidence-based research. The team calculated effect sizes for the meta-analysis in each intervention based on the available data, allowing a clearer understanding of the intervention effects. The currently available evidence shows that the majority of long-term, continuous exercise interventions have had significant positive effects on children and adolescents with autism, especially with respect to social skills. In addition to improving one's social life through regular exercise, benefits expand across other domains including improvements in motor performance, cognitive functioning, interpersonal relationships, behavioral management, general physical health, and brain function. These results are exciting and suggest exercise could be an important novel treatment for improving quality of life in people with autism.

The exercise interventions evaluated in the review included a range of types, from aerobic exercise, water exercise, and mind-body exercise such as yoga and Tai Chi, and structured ball sport. Nonetheless, the wide variation in exercise types, intensities, and durations made it difficult for the researchers to identify which approaches to physical activity—including, for example, which durations or frequencies—might provide the most benefits among this demographic. This highlights the essential need for more research to examine these factors in much greater detail. In general, the findings highlight an essential role of exercise as a potentially positive intervention strategy for children and adolescents with autism spectrum disorder, which can help improve their quality of outcomes and determinants for their development [28].

Lastly, according to a meta-analysis conducted by Li et al. [24] a network meta-analysis performed to evaluate and rank the efficacy of alternative physical activities to improve the respective functioning of children with autism spectrum disorder (ASD). To ensure an exhaustive collection of relevant literature, the researchers conducted a systematic search across several major databases including PubMed, EBSCO, Cochrane Library, and Web of Science until May 25, 2023. The analysis encompassed 37 studies and about 1200 participants, looking at 17 distinct physical interventions. Based on the ranking probabilities of all the activities, the study offered an indication that Tai Chi Chuan was potentially the most efficient intervention to improve motor functions. In addition, interventions focused on fundamental motor skills were identified as being potentially most beneficial for the improvement of communication skills in children with ASD.

Moreover, in assessing methods to tackle stereotyped behaviors and improve social functioning, kata techniques emerged as the most likely effective intervention. The study findings underscore the potential of physical activity as a key element in the comprehensive management of autism spectrum disorder, proposing that personalized physical interventions may promote notable advances in different functional areas for children impacted by the condition [24].

4. Sport and exercise for ADHD young's executive function

Attention-deficit hyperactivity disorder (ADHD) is classified as a neurodevelopmental and mental disorder characterized by symptoms such as disorganization and hyperactivity-impulsivity [15]. The global prevalence of ADHD is approximately 6%, with a higher incidence in boys compared to girls, specifically at a ratio of one girl for every three boys. This disorder adversely affects social interactions, academic performance, emotional well-being, and psychological health, leading to significant costs for both society and healthcare systems. Notably, around 60% of individuals diagnosed with ADHD continue to experience symptoms into adulthood, as sensory processing challenges may be associated with the ADHD phenotype [38].

A systematic review was conducted to examine the impact of physical activity, exercise, or sports on the executive functions of children and adolescents diagnosed with ADHD, utilizing existing scientific literature [38]. Research was considered for inclusion. The studies focused on interventions implemented for children and adolescents aged 5–18 years who have been diagnosed with ADHD. Exclusion criteria encompassed individuals younger than 5 years, those older than 18 years, athletes, and individuals diagnosed with other disorders or syndromes. Regarding the physical activity interventions, various programs were identified, including "exergames," activities focused on "coordination, balance, and acrobatics, among others," "cycle ergometer" sessions, "20-minute physical activity sessions," and "trampoline jumping." Each of these programs had a minimum duration ranging from 20 to 60 minutes per session. The exercise interventions also included programs utilizing the "cycle ergometer," "a combination of ball exercises, table tennis, and treadmill running," "moderate to vigorous and high-intensity exercises," "yoga," "aquatic exercises," "treadmill workouts," "treadmill training or whole-body vibration training," and "dynamic stretching, moderate-intensity interval training, along with perceptual-motor and dynamic games," with session durations ranging from 20 to 90 minutes. Lastly, the studies that implemented sports programs included "table tennis" and "taekwondo," with sessions lasting between 50 and 70 minutes.

This systematic review emphasizes the beneficial effects of physical activity on executive function across a range of programs, despite some limitations, particularly its focus on scientific research published up to August 2021 [38].

It evaluates the effectiveness of various intervention programs across different studies, leading to significant variations in the results obtained from the research examined (**Figure 2**).

Figure 2.
Executive functioning skills [13]. Available from: https://www.structural-learning.com/post/neurodevelopmental-disorders.

Engaging in physical activity yields beneficial outcomes for executive function, particularly through a combination of aerobic fitness and cognitively demanding exercises, as seen in a physical education program that included 2 hours of tennis-specific skill training [39].

Furthermore, physical exercise has shown to produce significant improvements in executive function, aligning with findings from a systematic review that suggests both acute and chronic exercise can enhance executive function. The review notes a small effect size for inhibitory control and cognitive flexibility, alongside moderate effect sizes for working memory [40]. Specifically, tennis-related interventions have demonstrated notable improvements in executive function, with one study concluding that table tennis enhances cognitive flexibility due to its demanding nature [41].

Although all interventions indicate enhancements in executive functions, there is considerable variability in the duration of each study. Furthermore, there is significant diversity in the types of physical activities, exercises, and sports programs examined, including yoga, running, table tennis, and swimming, among others. Additionally, the variations in the assessments used to evaluate executive functions complicate the ability to identify or recommend the most effective physical activity, exercise, or sports program for addressing ADHD symptoms, particularly in educational settings.

Potential influencing factors in the research include the high prevalence of the disorder among males, leading to very uniform samples regarding gender. Additionally, there is a lack of assessment or oversight concerning the physical fitness and sedentary habits of the participants in the interventions, as well as daily monitoring of the medications they use. Furthermore, the absence of follow-up after the interventions may cast doubt on whether the observed changes are substantial or merely temporary.

The authors suggest that future initiatives should include physical activity, exercise, or sports interventions of at least 20 minutes per session, in conjunction with a sports workshop, to improve executive functions in children and adolescents with ADHD. This strategy also promotes motor skills and social development, areas that are frequently impacted on individuals with ADHD.

In conclusion, the systematic review emphasized the positive impact of physical activity, exercise, and sports programs on executive function. Although there is a range of variability among these programs, they have shown significant enhancements in inhibitory control, processing speed, selective attention, cognitive flexibility, and various cognitive and behavioral functions, as well as improvements in academic performance and overall executive functions in children and adolescents with ADHD [38].

4.1 Social engagement as a fundamental component of exercise interventions for younger individuals with ADHD

Freire et al. [42] conducted a systematic review that included peer-reviewed studies published in English from 2000 to 2021, which aligned with the review's objectives. The methodological quality of 49 eligible studies—comprising 47 controlled trials and two mixed methods studies, with a total of 2355 participants—was evaluated using the mixed methods appraisal tool. Narrative synthesis revealed two categories of studies: (a) Group 1 included those with deliberate social participatory components and (b) Group 2 appeared to involve unintentional social participation. Most of the studies exhibited moderate-to-low methodological quality, and only a small number assessed the effects of interventions on overall physical activity levels.

The authors focused on synthesizing research related to exercise interventions that incorporated social participatory elements for children with complex developmental needs. High-quality studies that integrated social participation as a core component of exercise interventions described programs that included motor skills training or specific collaborative sports training, aimed at children aged 5–14 years [43].

Two studies, conducted by O'Connor et al. [3] and Pan et al. [44], employed the Bruininks–Oseretsky Test of Motor Proficiency (BOT-2) to evaluate participants' motor skills [3, 42], while Smith et al. [44, 45], utilized The Test of Gross Motor Development (TGMD-2). Pan et al. [44] reported significant enhancements in BOT-2 scores, cardiovascular endurance, and flexibility following their intervention, which featured a simulated developmental horse-riding program, ball games, and exercise stations, concluding with cool-down activities that emphasized "prosocial behavior and engagement." However, Pan et al. [44] focused solely on physical outcome measures, despite aiming to provide an intervention suitable for children with ADHD that included social components [42].

O'Connor et al. [3] assessed the impact of sports training within an intensive behavioral modification program for children with ADHD, discovering notable improvements in both motor and sports skills, including performance and understanding of game rules. They requested parents to evaluate their child's progress in "sports skills" and "good sportsmanship" using a Likert scale and found that parents of children in the intervention group reported significantly greater improvements compared to those in the control group [3].

Additionally, Chou and Huang [46] reported significant advancements in attention and discrimination functions in children with ADHD following an 8-week yoga program [44].

The analysis conducted by Freire et al. [42] identified short-term enhancements in various physical outcomes, particularly motor skills, following exercise interventions that included social participation [42].

However, the long-term effects of these improvements remain unexplored. Future research should adopt longitudinal designs to assess the sustainability of both physical and psychosocial outcomes. Additionally, while short-term gains in motor skills were observed, many children still did not achieve their developmental motor milestones [42]. This indicates that future research should not only focus on longitudinal studies but also incorporate repeated interventions or booster sessions [45]. Booster sessions are now recognized as a valuable element of behavior change strategies, helping to reduce the decline in motivation and routine that can lead to setbacks [45]. Understanding both the persistence of outcomes over time and the effects of booster sessions would equip clinicians and families with valuable insights for planning effective interventions.

In the systematic review conducted by Freire et al. [42] in Group Two, high-quality research suggested that social participation was likely a secondary element of the group environment. The studies included exercise interventions that featured motor skills training (n = 4, 50%) and obstacle courses (n = 3, 37.5%). Among the high-quality studies in this group, six (76%) focused on a specific area of participant development: Four (50%) aimed at physical outcomes, one (12.5%) addressed psychological outcomes, and another (12.5%) investigated executive function outcomes. Of the five studies in Group Two that assessed motor skills, three (60%) utilized the Movement Assessment Battery for Children-2, with all five studies reporting significant improvements in motor skills. Furthermore, three (37.5%) studies employed different physical outcome measures, including research by Fong et al. [43], which revealed significant enhancements in sensory organization and unilateral standing tests among 16 children with DCD after 3 months of Taekwondo training. However, none of the high-quality studies in Group Two assessed the overall physical activity post-intervention levels.

It was a plausible assumption that high-quality studies incorporating social elements into exercise interventions would yield more significant improvements across various domains compared to those where social participation was merely incidental. However, there is a scarcity of studies evaluating outcomes across multiple domains, resulting in limited understanding of how group exercise interventions affect the social domain and other facets of a child's development [43].

Nonetheless, the review has pinpointed several areas for further exploration, such as sports skills and sportsmanship [3]. Improvements in physical outcomes, especially in motor skills, were commonly observed and identified as the primary advantage of social exercise programs for children with complex developmental needs, where evidence is available. However, further comprehensive and long-term studies are necessary to evaluate the social, psychological, and executive function effects of these interventions within this population. Additionally, these programs should include booster sessions to enhance children's chances of achieving developmental milestones. It is crucial to conduct further research to identify the contexts that positively and negatively influence these outcomes within the population with complex developmental needs (such as ADHD, DCD, ID, ASD) [43].

An innovative study by Kambas et al. [47] investigated the effects of group psychomotor therapy (GPT), a program focused on physical activity and inclusion, on the social-emotional and motor skills of 54 preschool children, including 19 children diagnosed with ADHD and 35 children without ADHD (Mage = 61.80 ± 5.40 months). These children participated in a 27-week GPT program. Previous research has shown that physical activity initiatives can benefit children with ADHD by enhancing neuropsychological and cognitive functions, executive skills, and social-emotional development. The assessment of motor and social-emotional skills was conducted before and after the intervention program using the DEMOST-PRE© and the Psychosocial Adjustment Test, respectively. Findings revealed that while children without ADHD outperformed their peers with ADHD in all measured domains, both groups showed significant gains in motor and socio-emotional skills. While the lack of a control group presents a limitation in this study, the findings highlight the effectiveness of GPT as a beneficial practice that can enhance other therapeutic methods. Additionally, the results countered the belief that it is unfeasible to include both children with and without ADHD in a group environment. The authors emphasized that the benefits of these interventions are not exclusive to children with ADHD [46].

4.2 Concluding observations and suggestions for children and adolescents with ADHD

Research indicated that physical activities positively influence motor skills, cognitive abilities, and overall functioning in individuals with ADHD [48]. Specifically, aerobic exercises (e.g., running, cycling, swimming, table tennis, coordination exercises), which enhance blood circulation to the brain, were found to significantly improve executive functions and verbal working memory in these individuals. Further research is needed to explore the long-term advantages of aerobic exercise, as it has been recognized as a valuable component in a comprehensive strategy for managing ADHD symptoms. Aerobic exercises promote neuroplasticity, which is the brain's capacity to reorganize and form new connections. This is especially beneficial for individuals with ADHD, who often face challenges with executive functions. Engaging in regular physical activity not only boosts physical fitness but also enhances cognitive and emotional capabilities, leading to improved overall performance. In summary, aerobic activities serve as valuable support for those with ADHD, aiding in the development of self-regulation and concentration skills while also contributing to mental and physical well-being [47].

Marwaha et al. [49] investigated the advantages that aquatic exercise can offer in addressing key motor and behavioral symptoms of the disorder, including hyperactivity and impulsiveness, and its subsequent impact on enhancing psychological and emotional health, as well as the social integration of children. A 12-week interventional sports program was designed for 40 children aged 8–12 who have been diagnosed with ADHD and various conditions. To assess the advantages of swimming, a model involving both an experimental group and a control group was utilized. The experimental group engaged in the program, which included three low-intensity training sessions each week, lasting 90 minutes per session, while the control group continued with their regular schedule. The results indicated not only advancements in sports performance and physical fitness but also a notable enhancement in the children's concentration and attention span. This led to a decrease in hyperactivity and impulsive behaviors, with scores improving from 17.12 to 9.76. Furthermore, when compared to the control group, there was a marked reduction in nervousness and an increase in cooperative and team-oriented behaviors, with scores changing from 10.72 to 6.81. This improvement can be attributed to their involvement in a structured and organized activity. Additionally, several psychological benefits were observed, including an increase in self-esteem, which contributed to lower levels of anxiety and emotional instability, facilitating better social integration [50].

Regarding martial arts (e.g., Judo, Taekwondo, Capoeira), they serve as a beneficial supplement for enhancing children's attention and concentration. This form of activity also aids in managing hyperactivity by providing a creative outlet [51].

Yoga practices represent another effective method for assisting children with ADHD, providing a gentle and natural way to enhance their quality of life by better managing anxiety [49, 52–54]. In the research conducted by Cohen et al. [54], the focus was on the use of mindfulness and meditation techniques for managing ADHD symptoms in children and adolescents. This review draws upon published clinical data and studies that have implemented yoga, mindfulness, or meditation practices for this demographic. The findings indicated a positive impact on symptom reduction, particularly in enhancing attention and decreasing frequent outbursts and impulsivity. Participants also reported improved emotional regulation and coping strategies for anxiety. However, it is important to acknowledge certain limitations in the studies and meta-analyses, including the diversity of implementation methods and the short duration of most programs. The authors recommended further exploring the impact of various yoga styles and other alternative non-pharmacological interventions on symptom management. Also, there is a need for larger and more extended studies to validate these results and to gain a deeper understanding of the long-term effects [52, 53].

As a result, participating in playful activities, especially aerobic exercises, has been proven to greatly improve social skills, which in turn enhances relationships with peers and helps to regulate intense behaviors. In conclusion, every type of activity has shown a positive and diverse effect on the functioning of individuals with ADHD.

Thus, it is crucial to conduct further research that highlights the therapeutic benefits of intervention programs, focusing on larger sample sizes and longer follow-up durations. Therefore, relying solely on medication is inadequate without the advantages gained from physical activities, as their combination results in more effective management of the disorder, addressing both cognitive-motor and psychological dimensions [53].

5. Sport and exercise for DCD young's executive function

The DSM-5 defines Developmental Coordination Disorder (DCD) as a condition that significantly hinders the development of motor coordination skills, resulting in various difficulties for children and adolescents in their daily activities [15].

The prevalence of DCD in children varies between 2 and 20%, with the most cited figure in the literature being between 5 and 6% [54]. Typically, children with DCD execute motor tasks, such as walking and writing, at levels considerably below what is expected for their age group. These difficulties often lead to a marked decline in academic performance and the ability to engage in everyday tasks. Moreover, the occurrence of comorbid conditions is significantly elevated in children diagnosed with generalized perceptual-motor disorders [55]. Developmental Coordination Disorder (DCD) is commonly associated with attention hyperactivity disorders/ADHD (35–50%), specific language impairments (SLI), dyslexia (35%), dyscalculia, autism spectrum disorder (ASD), Asperger's syndrome, and Tourette's syndrome, in addition to issues related to anxiety, depression, and challenges in social interactions and self-esteem (see **Figure 3**). In clinical settings involving DCD, the presence of comorbidities is typically more prevalent than not [56]. These co-existing conditions can increase the tendency to avoid physical activities. The repercussions of DCD include diminished participation in team sports, lower self-esteem, emotional or behavioral issues, impaired academic success, poor physical fitness, reduced physical activity, and obesity. Importantly, these movement-related difficulties are not attributable to intellectual disability or neurological conditions, such as cerebral palsy [15].

The literature review highlights deficiencies in both fine and gross motor skills, as well as in poor visual perceptual skills and executive functions (EFs)—which encompass planning, task switching, response inhibition, and memory storage—as two key features of DCD [56–60].

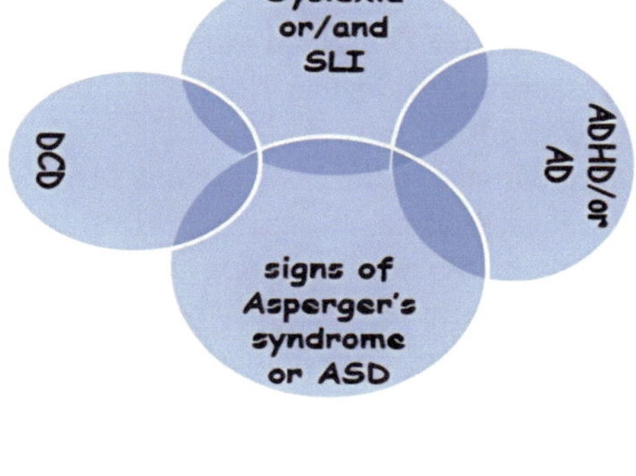

Figure 3.
Comorbidity of DCD [56].

The research conducted by Fujisawa et al. [61] aimed to explore the connection between developmental coordination disorder (DCD) traits in preschoolers and their emotional and behavioral issues, while accounting for the influence of autistic traits. The findings indicated that autistic traits correlate with emotional and behavioral challenges in preschool children. However, even after adjusting for these autistic traits, a significant association remained between higher DCD traits and increased conduct problems, hyperactivity/inattention, and peer relationship difficulties, along with a decrease in prosocial behavior. This underscores the importance of providing support for children facing motor skill challenges. Addressing DCD traits not only enhances motor abilities but also helps in reducing emotional and behavioral issues among preschoolers [62].

The study by Van Dyck et al. [23]—comprising 50 children with DCD and 31 typically developing peers—revealed that a significant proportion of children aged 7–11 with DCD struggled with multiple executive functions (92%) and praxis (68% for non-functional gestures and postures), and also exhibited difficulties in attention (52%), visual perceptual skills (46%), and visuomotor skills (36%).The clustering analysis revealed five distinct subtypes: (i) generalized impairments (8 children with DCD), (ii) difficulties in manual dexterity, balance (both static and dynamic), planning, and alertness (15 children with DCD and 1 typically developing child), (iii) challenges in manual dexterity, cognitive inhibition, and visual perception (11 children with DCD), (iv) problems with manual dexterity and cognitive inhibition (15 children with DCD and 5 typically developing children), and (v) no impairments (25 typically developing children and 1 child with DCD). Although there were slight variations, the motor and praxis assessments did not effectively distinguish between the four subtypes of children with DCD. Nevertheless, each subtype was characterized by specific perceptual or cognitive difficulties. These results highlight the necessity of a comprehensive evaluation of the perceptual and cognitive skills in children with DCD, given the cognitive dysfunction of DCD profile [61].

Asonitou and Koutsouki's recent study [57] investigated the significant role of executive dysfunction as a common secondary characteristic associated with movement disorders. Internal speech is crucial for organizing motor tasks, acting as a silent bridge between cognition and action [63]. The research identified balance, overall motor scores, and planned codes as key factors differentiating preschool children with and without DCD. A strong Pearson's correlation was found between the total scores on the MABC-2 and the subdomain scores, as well as the Planning Scale indices for both groups, indicating a meaningful connection between these evaluations. Among the items related to Planned codes, one emerged as the most significant predictor of overall motor performance in children. Additionally, the total score for Planned Connections was recognized as the strongest predictor for tasks requiring manual dexterity. In conclusion, the interplay between motor skills and executive functions during early childhood is essential for developing early intervention strategies that incorporate cognitive-motor tools (**Figure 4**) [56].

An alternative perspective investigated how various forms of exercise impact a child's cognitive abilities. For instance, allowing children to run a lap around the track before a one-hour math exam could enhance their focus. Research indicates that even a brief 20-minute walk can significantly boost children's attention span, overall executive function, and their performance in math and reading assessments. Additionally, a vigorous walk may assist children with attention-deficit hyperactivity disorder in improving their concentration, although the duration of these benefits remains uncertain [65].

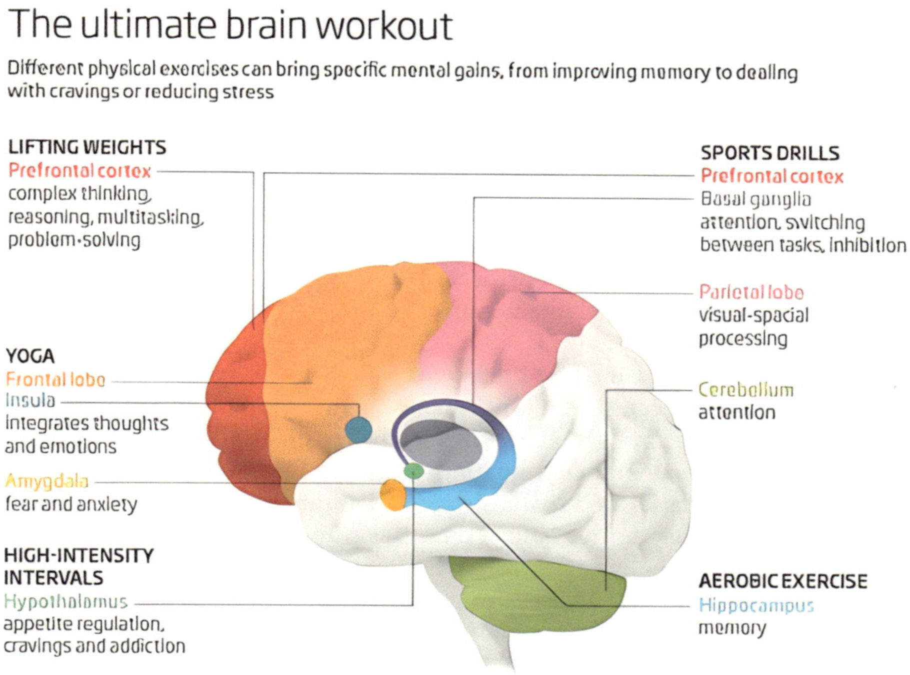

Figure 4.
Training in relation to the brain. Available from: https://inalliancepse.org/new-finding-different-types-of-exercise-affect-different-parts-of-your-brain/ [64].

Hillman et al. [66] argued that children should participate in at least 1 hour of physical activity daily after school. They highlight the importance of spreading this exercise throughout the day for maximum benefit. Engaging in aerobic activities aids children in maintaining focus on their goals. Furthermore, incorporating movement or walking breaks every 2 hours is considered more effective for enhancing learning, as it provides essential pauses. In contrast, structured exercises designed to develop specific skills, such as sports or coordination, may hinder attention [64]. Hillman's research focused on children aged 8–11, a period when crucial brain areas like the hippocampus and basal ganglia are still maturing. Thus, early intervention can have a profound effect on their development. Even slight improvements in fitness can result in noticeable changes in brain function. Numerous studies involving children in after-school physical activity programs have consistently demonstrated cognitive benefits. In many instances, children with obesity may not see significant weight loss, yet their cognitive function improves. Overall, encouraging fitness in children can lead to enduring cognitive benefits [34].

5.1 Social engagement as a fundamental component of exercise interventions for younger individuals with DCD

Research indicates that children with DCD engage in physical activities less frequently than their peers without the disorder, primarily due to their lower motor skills and negative evaluations of their performance from parents, teachers, and peers. Consequently, they often shy away from team sports, leading to social isolation and hindering the development of essential interpersonal skills. Motor competence is acknowledged as a key factor influencing children's social standing among peers. Children with DCD frequently experience low social status, which can manifest in scenarios such as not being chosen or being the last selected for sports teams. This avoidance of team games and physical activities can further diminish their perceived competence and negative impact on their motor skills due to insufficient practice (**Figure 5**) [66].

Interventions aimed at addressing DCD in children and adolescents exhibit considerable diversity, each targeting various aspects of DCD, encompassing both physical and psychological dimensions. Furthermore, these interventions yield beneficial effects on cognitive skills, social interactions, emotional well-being, and engagement in recreational sports activities [67].

The application of these interventions across different age groups, including both children and adolescents, results in enhancements in motor skills—such as locomotion (walking, running), manual dexterity, dynamic and static balance—as well as improvements in motor coordination, gross and fine motor abilities, and ambidextrous manipulation and spatial awareness. A range of exercise modalities has been employed, including team training, dual-task training, game-based activities, therapeutic riding, trampoline exercises, treadmill workouts, balance beam activities, and swimming [68, 69].

Zaragas et al. [70] conducted a comprehensive review of research conducted from 2014 to 2022, focusing on various physical activity interventions, particularly their duration and frequency, for children and adolescents aged 4–18 years who have been diagnosed exclusively with DCD and associated conditions, including ADHD, Down syndrome, and hearing impairments. This review emphasizes research on a variety of sports, including martial arts, aerobic workouts, swimming, and trampoline activities. Also, it indicates that there are intervention programs incorporating physical activity for children and adolescents with DCD that effectively enhance their motor skills. Beyond the improvements in motor abilities, these physical activity interventions also influence the social and emotional growth of these individuals, with effects that can be both beneficial and detrimental. Notably, aerobic exercise has been found to alleviate anxiety and depression levels, while also enhancing quality of life and academic achievement. Furthermore, game-based physical education programs have demonstrated improvements in attention, organization, self-control, planning, decision-making, executive functions, and self-efficacy [71].

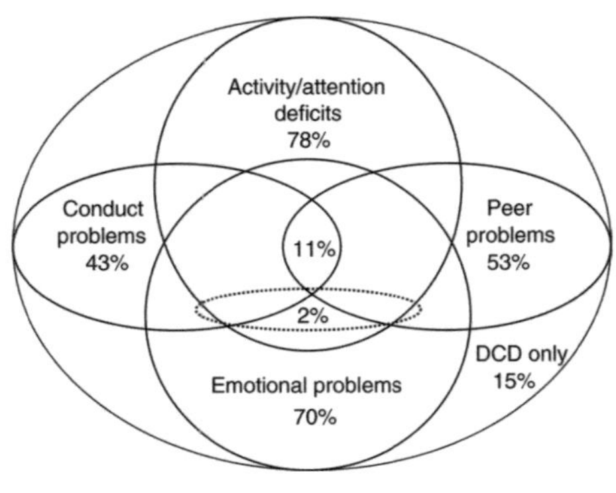

Figure 5.
Overlap between emotional and behavioral disorders in children with developmental coordination disorder (DCD) [54].

Gross mobility programs seem to foster feelings of enjoyment and satisfaction among participants, enhancing attention, cooperation, engagement in moderate to vigorous physical activity, self-efficacy, and ultimately improving parents' perceptions of their child's physical activity performance. In programs that included trunk stabilization exercises, there were noticeable improvements in certain quality-of-life health factors and functional status of the participants. However, findings related to Body Mass Index (BMI) and trunk stabilization exercises are inconsistent; one study indicates a significant reduction in BMI, while another finds no notable changes [70].

Furthermore, both high- and low-intensity physical activity programs that elevated metabolic rates were found to decrease adolescent fatigue after sessions, although they increased fatigue during the activities. Two interventions were carried out during the camp aimed at enhancing self-efficacy, self-confidence, attention, organization, and communication skills, along with encouraging participation in extracurricular activities. Furthermore, a program that incorporated horse riding contributed to the development of social and emotional skills, behavior, and cognitive abilities among the participants. Following an intervention that focused on fundamental motor skills, children exhibited a reduction in sleep disturbances [71].

In a program centered on task-oriented activities and motor skills training, children displayed increased anxiety and decreased enjoyment after the intervention, despite parents observing improvements in rhythm functioning and a decrease in peer-related issues [71]. Interestingly, when children were allowed to select the skills, they wished to develop, their anxiety levels appeared to diminish [72]. Quiet Eye interventions have been effective in enhancing attention and concentration in children with DCD, and parents have reported improvements in their children's self-confidence, social skills, and willingness to engage in physical activities post-intervention. Additionally, dual-task balance exercises have been found to enhance cognitive abilities [71].

The authors concluded that formal education, beginning at the kindergarten level, should emphasize Physical and Psychomotor Education Programs. They highlighted the need for additional training for kindergarten and physical education teachers to successfully implement treatment programs for DCD through specific physical activities. It is vital to concentrate on engaging and appropriate physical activities that cater to the developmental needs and interests of children, while also equipping educators with the training required to effectively manage DCD.

Monastiridi et al. [67] performed a systematic review of 22 studies, emphasizing that the fitness levels of children and adolescents with DCD are essential for their everyday functioning and have a beneficial impact on their health-related quality of life (HRQOL). Research indicates that all aspects of HRQOL—motor, cognitive, emotional, and social—are significantly lower in children and adolescents with DCD compared to their peers. This underscores the necessity of interventions aimed at enhancing motor skills and HRQOL in this population. This study sought to systematically review literature from peer-reviewed journals to summarize the potential connections between intervention strategies focused on physical fitness and exercise, participation in physical activities, and HRQOL among children and adolescents with DCD. It also included studies that assessed the impact of fitness and exercise intervention programs on motor competence and HRQOL in this population.

This literature review revealed that various intervention programs tailored for individuals with DCD, aimed at improving motor abilities, have been developed from fields such as occupational therapy, physiotherapy, medicine, dietetics, and education. Overall, the review concentrated on two primary approaches in clinical practice for DCD interventions: *the task-oriented approach*, which focuses on enhancing specific skill performance, and *the process-oriented approach*, which seeks to identify and address the underlying processes or dysfunctions that have not been adequately developed for the individual's age yet are essential for effective motor skill performance and acquisition.

Children and adolescents diagnosed with DCD appear to experience lower health-related quality of life (HRQOL) compared to their typically developing counterparts. However, further research is necessary to fully understand HRQOL in this population. There is a necessity for studies on interventions that emphasize fitness and exercise programs aimed at enhancing motor skills and HRQOL. By participating in these interventions, children and adolescents with DCD may have the opportunity to disrupt the negative cycle of avoiding physical activity and instead foster a more positive engagement. It is suggested that there may be beneficial connections between fitness and exercise programs, motor competence, and HRQOL among children and adolescents with DCD. Future studies should investigate how these interventions can alleviate functional limitations, encouraging participation in physical activity and subsequently improving HRQOL within the DCD community (**Figure 6**) [66].

The links between poor motor coordination and previously identified psychosocial factors indicate that social and environmental risk and protective elements play a crucial role in the causal chain. Cairney et al. [73] and Mancini et al. [74] have proposed "the environmental stress hypothesis" through a series of publications, providing a theoretical framework to explore the connections among DCD, social and

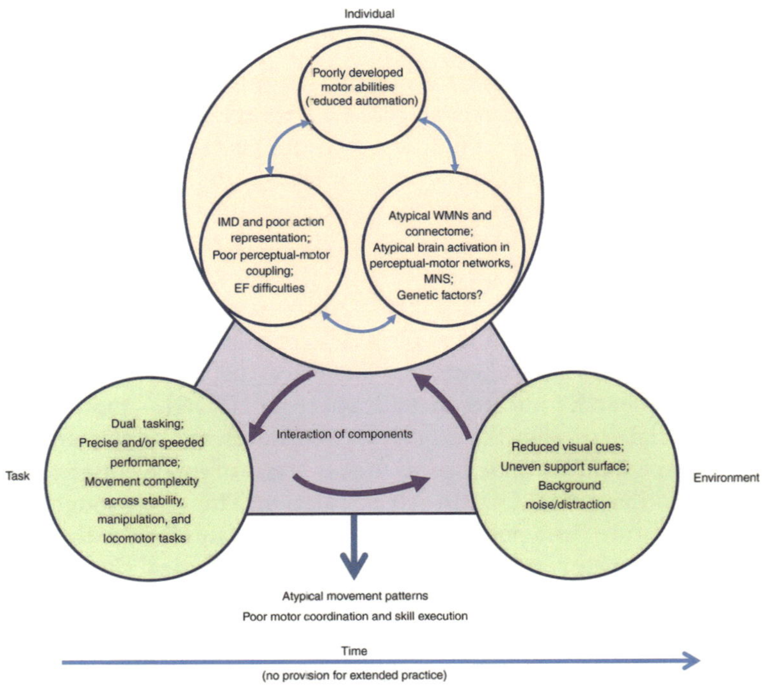

Figure 6.
A comprehensive overview of motor skill development highlighting the factors associated with performance in DCD. Key components include Internal Modeling Deficit (IMD), Executive Function (EF), White Matter Network (WMN), and Mirror Neuron System (MNS) [54].

individual factors, and mental health. Utilizing the stress process framework, DCD or poor motor coordination is identified as a primary stressor, which subsequently subjects children and adolescents to various secondary stressors stemming from both interpersonal and intrapersonal conflicts, such as bullying and daily challenges related to functional limitations and self-perception. These stressors can directly and indirectly influence symptoms of internalizing issues such as depression and anxiety. Indirectly, stress linked to motor coordination difficulties may be mitigated by supportive factors, such as caring parents or a sense of competence [72, 75]. A particularly innovative aspect of this model is the incorporation of overweight/obesity and physical inactivity as pathways that connect DCD to mental health issues by affecting stress levels and psychosocial resources [54].

Research indicates that both inadequate physical activity and unhealthy weight are prevalent and associated with mental health challenges in this population. Several studies have examined components of this model, with the most robust evidence highlighting self-concept (including self-esteem and self-worth) as a mediator in the relationship between DCD and mental health issues in children and adolescents. Harrowel et al.'s study [76] specifically identified low self-esteem and poor social communication as factors linked to mental health and well-being challenges [75]. However, a thorough evaluation of the entire model has yet to be undertaken. Furthermore, a significant limitation identified in the reviewed literature is the inconsistency among studies, along with a sometimes-inadequate evaluation of motor difficulties. In certain cases, the evidence supporting the existence of substantial motor challenges is relatively weak [54]. The authors suggest that longitudinal research is essential to gather further insights into the developmental trajectory of DCD during adolescence and adulthood. Additionally, there is a need for age-appropriate standardized assessments to facilitate the diagnosis and evaluation of interventions for DCD in these age groups.

The European Academy of Childhood Disability (EACD) has established 35 clinical practice recommendations (CPR-DCD) for DCD. Among these, eight are derived from literature reviews, three focusing on "assessment" and five on "intervention." Additionally, 22 recommendations have been revised from the 2012 guidelines. New insights include two good clinical practices (GCPs) related to diagnosis and assessment, as well as three GCPs addressing psychosocial concerns. Furthermore, a new recommendation (level of evidence) highlights the use of active video games as supplementary tools alongside traditional activity and participation-focused interventions. Two additional recommendations (one GCP and one level of evidence) have been introduced specifically for adolescents and adults with DCD. The CPR-DCD provides a vital and comprehensive overview of DCD, integrating research findings and expert consensus, and serves as a valuable resource for clinicians and researchers across various fields. These international guidelines may also inform the development of national standards [54].

A recent study on digital technology by Smits-Engelsman et al. [77] evaluated the effects of graded exergames on children aged 7 to 12 with DCD in comparison with their typically developing (TD) peers. A total of 60 participants, consisting of 32 children with DCD and 28 TD children, engaged in 30-minute training sessions twice a week over a span of 10 weeks. Assessments of motor coordination (using the MABC-2 test), balance, and both aerobic and anaerobic fitness were conducted at the start and conclusion of the training program. Additionally, participants' enjoyment and perceived exertion levels were recorded throughout the training. Results showed that both the DCD and TD groups made significant gains in motor coordination, balance, and fitness by the end of the training. A notable interaction between Group and Time was found in the MABC-2 total scores [$F(1,55) = 13.19$; $p < 0.001$] and balance scores [$F(1,55) = 26.83$; $p < 0.0001$], with the DCD group exhibiting greater improvements than their TD peers. Despite rating the training as high in intensity, both groups reported enjoying the program. These results suggest that graded exergames could be an effective intervention for enhancing fitness in children with DCD. To improve and sustain fitness levels, regular engagement in graded exergames within school environments may be beneficial for both children with and without DCD [69].

5.2 The impact of new technologies as intervention tools for individuals with DCD and ADHD

The review conducted by Pergantis [78] highlighted how the incorporation of Information and Communication Technologies (ICT) into education fosters the inclusion of students with special educational needs (SEN) and disabilities. This integration enables these students to articulate their thoughts in diverse ways, both verbally and non-verbally, while stimulating their creativity. Furthermore, educators exhibit a supportive attitude toward ICT integration, acknowledging its crucial role in facilitating equal participation among all students, including those with SEN or disabilities. By leveraging technology in the classroom, students become more engaged and active participants rather than passive observers, which enhances their sense of autonomy and accountability for their learning journey. In a virtual reality (VR) learning environment, students can practice social skills in a safe, varied, and motivating setting that is both holistic and realistic. Gamification in education serves as a method to boost engagement by integrating game-like elements into the learning experience.

Students with ADHD can gain from computer-assisted learning, especially through game-based approaches, which can lead to improvements in their behavior, academic performance, and self-regulation. Research indicates that gaming significantly increases engagement and motivation among individuals with ADHD. Active learning provides these students with opportunities to develop their skills in an environment tailored to their specific needs and potential.

Ultimately, the application of ICT transcends mere knowledge acquisition; it enhances students' motivation to learn and fosters interaction within the educational setting, thereby promoting self-esteem and a sense of achievement. The symptoms commonly associated with children with neurodevelopmental disorders (NDDs) can greatly affect their integration into school life. The findings of this study suggest that the implementation of new technologies, such as ICT, may positively influence the participation and engagement of the studied populations within the school environment [77].

Also, the utilization of new technologies as intervention tools was highlighted, underscoring the crucial role of digital technologies in education and DCD training [78]. These technologies are not only effective but also enhance assessment, intervention, and educational processes through mobile devices, enabling educational activities to be conducted anywhere. Various ICT applications serve as key supporters of education, while advancements in AI, STEM, and robotics elevate educational practices to new heights. Additionally, engaging games contribute positively to learning experiences. The integration of ICT with theories and models related to metacognition, mindfulness, meditation, and emotional intelligence development significantly accelerates and enhances educational outcomes, particularly for children with DCD.

Several studies have focused on these three main categories of technology, showing positive results in addressing various challenges related to ADHD, such as motor skills, visuomotor integration, and academic achievement. Although most studies showed beneficial effects on skill development and overall participant outcomes, many faced limitations, such as small sample sizes, short intervention periods, and methodological issues, which affect the generalizability of the findings. Further research is needed to empower practitioners and healthcare providers to effectively integrate these tools into treatment plans and daily activities.

Wearable technologies are a significant tool in information and communication technology (ICT) as they facilitate the monitoring of neurophysiological changes and heightened activation of the hypothalamic-pituitary-adrenal (HPA) axis. This activation is often observed in individuals with ASD when they face non-threatening environmental stimuli or social interactions that may induce stress [78].

In a study by Păsărelu et al. [79], the effectiveness of mobile applications in alleviating symptoms of ADHD was systematically evaluated. The researchers reviewed resources from iTunes/iOS (Apple App Store), Google Play, and the National Health Service Health App Library between May 2017 and September 2019. They focused on apps specifically designed for ADHD that addressed goal assessment, treatment, or both, ensuring they were functional and available in English. Out of an initial pool of 355 apps, only 109 met the criteria for inclusion. The findings indicated that very few of these apps provided information about their development process, and none offered evidence supporting their effectiveness. Research involving tablets has yielded promising results for children with DCD, demonstrating their effectiveness across various settings, including classrooms [80].

For instance, Coutinho et al. [81] conducted an experimental pilot study that examined the impact of iPad applications on enhancing visual-motor skills in children with special needs, specifically those aged from 4 years and 11 months to 7 years and 11 months. The study revealed significant improvements in the visuomotor integration skills of the participants [79].

For children with ADHD, particularly those facing challenges with executive functions, studies have shown that game-based interventions can lead to positive outcomes, enhancing working memory, inhibitory control, planning, and programming skills. In the case of DCD, most interventions have primarily focused on addressing deficits in motor skills, academic performance, and physical fitness, with encouraging evidence supporting the use of tablets, applications, virtual reality (VR), and active video games (AVGs) [77, 81].

In the study conducted by Purpura et al. [82], children with DCD were assessed before and after an 11-week virtual reality (VR) training program. The evaluation focused on various aspects, including visual attention, inhibition, planning skills, and visual-motor coordination. The rehabilitation programs were tailored to meet the specific clinical needs and functional profiles of each child, incorporating a range of games with differing levels of complexity. The initial results indicated a significant overall improvement in executive functions, particularly in visual attention and inhibition abilities. These outcomes highlight important implications for clinical practice, offering valuable insights for professionals on the use of VR in pediatric rehabilitation [83].

Conclusively, the research findings indicate that numerous technologies have been implemented and evaluated within educational environments, demonstrating beneficial impacts on the engagement of students with neurodevelopmental disorders (NDDs).

6. Concluding observations and suggestions

Engaging in physical activity (PA) offers a valuable opportunity to improve mental health in children and adolescents who have neurodevelopmental disorders. However, the specific mechanisms at play are not yet fully understood. The systematic review conducted by Liu et al. [40] focused on the mechanisms through which PA interventions may influence mental health in children and adolescents with NDDs. In total, the review included 45 studies, comprising 29 randomized controlled trials and 16 quasi-experimental studies, with a combined participant count of 1751. According to the best evidence synthesis rating system, the most frequently examined and consistent mechanisms through which PA influenced mental health in children and adolescents with NDDs were neurobiological (theta activity[1] and P3 amplitude[2]), psychosocial (social skills and social participation), and behavioral (motor skills and sleep). However, the evidence regarding P3 latency, beta-activity,[3] and physical self-concept was found to be insufficient. In conclusion, future PA interventions should consider the neurobiological, psychosocial, and behavioral mechanisms identified. PA can be developed as a supplementary approach alongside interventions that specifically target these mechanisms to enhance mental health in children and adolescents with NDDs [40].

Systematic reviews have determined that self-esteem, self-concept, and self-efficacy are the primary pathways through which physical activity (PA) influences both internalizing and externalizing issues throughout the lifespan. Additionally, it has been suggested that neurobiological factors (such as hippocampal volume and brain structure) and behavioral aspects (like sleep) may serve as mechanisms by which PA impacts cognitive function across different age groups. A conceptual framework has been introduced to clarify three possible mechanisms linking physical activity (PA) interventions to mental health outcomes in typically developing youth: neurobiological factors (including gray matter volume and activation), psychosocial elements (such as physical self-concept and social connectedness), and behavioral aspects (covering sleep quality and self-regulation skills). Research indicates that brain structure, physical self-concept, and self-regulation are the most examined and reliable mechanisms identified in young people [34].

Interventions involving physical activity and sports for children and adolescents with NDDs should be tailored to accommodate the severity of their conditions and their specific needs. Inclusive education is a fundamental objective of the UN Convention on the Rights of Persons with Disabilities (CRPD) and various initiatives led by the UN and UNESCO. Within this context, researchers have been investigating how sports and physical education can facilitate social inclusion and integration. Findings reveal that individuals of all genders face different types of social exclusion—such as attitudinal, environmental, and institutional—but engaging in sports greatly improves their quality of life. The research [84] suggests that sports help individuals develop a social identity,

1 *Theta activity:* refers to an increase in the power of the EEG theta band, which has been associated with treatment outcomes in individuals suffering from major depressive disorder (MDD).

2 *P3 amplitude:* parietal P3 amplitude of the Event-Related Potential (ERP). The P300 (P3) wave is an ERP that arises during decision-making processes. It is classified as an endogenous potential because its manifestation is associated not with the physical characteristics of a stimulus, but rather with an individual's response to it.

3 *Beta-activity*: Beta-waves are characterized by their high frequency and low amplitude, typically present when a person is awake. These brain waves play a crucial role in conscious thought and logical reasoning, often providing a stimulating effect. An optimal level of beta-waves is essential for maintaining focus.

a sporting identity, and a sense of purpose, while also promoting their social integration and inclusion. Many obstacles arise from a lack of awareness, information, and understanding regarding both the social and biological dimensions of disability. Sports have the potential to challenge and change prevailing stereotypes and attitudes.

Author details

Katerina Asonitou*, Sophia Charitou* and Dimitra Koutsouki
Laboratory of Adapted Physical Activity/Developmental and Physical Disabilities, School of Physical Education and Sport Science, National and Kapodistrian University of Athens, Athens, Greece

*Address all correspondence to: kasonitou@phed.uoa.gr and sofhar@phed.uoa.gr

References

[1] Morris-Rosendahl DJ, Crocq M-A. Neurodevelopmental disorders-the history and future of a diagnostic concept. Dialogues in Clinical Neuroscience. 2020;**22**(1):65-72. DOI: 10.31887/DCNS.2020.22.1/macrocq

[2] Lubans D et al. Physical activity for cognitive and mental health in youth: A systematic review of mechanisms. Pediatrics. 2016;**138**(3):e20161642. DOI: 10.1542/peds.2016-1642

[3] O'Connor BC et al. Effects of a summer treatment program on functional sports outcomes in young children with ADHD. Journal of Abnormal Child Psychology. 2014;**42**(6):1005-1017. DOI: 10.1007/s10802-013-9830-0

[4] Bull FC et al. World Health Organization 2020 guidelines on physical activity and sedentary behaviour. British Journal of Sports Medicine. 2020;**54**(24):1451-1462. DOI: 10.1136/bjsports-2020-102955

[5] Lubans DR et al. Fundamental movement skills in children and adolescents: Review of associated health benefits. Sports Medicine (Auckland, N.Z.). 2010;**40**(12):1019-1035. DOI: 10.2165/11536850-000000000-00000

[6] Ferguson GD et al. Health promotion in a low-income primary school: Children with and without DCD benefit, but differently. Physical and Occupational Therapy in Pediatrics. 2015;**35**(2):147-162. DOI: 10.3109/01942638.2015.1009230

[7] Zaccaria V et al. Multiple complex developmental disorder (MCDD): Did we throw the baby out with the bathwater too fast? A systematic review. Clinical Child Psychology and Psychiatry. 2025;**30**(1):5-19. DOI: 10.1177/13591045241285486

[8] McWhirter K et al. The association between learning disorders, motor function, and primitive reflexes in pre-school children: A systematic review. Journal of Child Health Care: For Professionals Working with Children in the Hospital and Community. 2024;**28**(2):402-428. DOI: 10.1177/13674935221114187

[9] Bremer E et al. Effectiveness of a fundamental motor skill intervention for 4-year-old children with autism spectrum disorder: A pilot study. Autism: The International Journal of Research and Practice. 2015;**19**(8):980-991. DOI: 10.1177/1362361314557548

[10] Syropoulou A et al. The presence, enjoyment, mood experience, attitude and preference towards exergames scale in children with mild intellectual disability. Sport Sciences for Health. 2022;**18**(1):203-210

[11] Choo YY et al. Developmental delay: Identification and management at primary care level. Singapore Medical Journal. 2019;**60**(3):119-123. DOI: 10.11622/smedj.2019025

[12] Piek JP et al. Motor coordination and social–emotional behaviour in preschool-aged children. International Journal of Disability, Development and Education. 2008;**55**(2):143-151

[13] Gawish A. Neurodevelopmental Disorders. 2023. Available from: https://www.structural-learning.com/post/neurodevelopmental-disorders

[14] Dasso NA. How is exercise different from physical activity? A concept

analysis. Nursing Forum. 2019;**54**(1): 45-52. DOI: 10.1111/nuf.12296

[15] American Psychiatric Association, editor. Diagnostic and Statistical Manual of Mental Disorders: DSM-5. 5th ed. American Psychiatric Association: Washington, DC, USA; 2013

[16] Diz S et al. The effects of a multi-sports program on the physical fitness, quality of life, and well-being of people with intellectual and developmental disabilities—A study protocol. Disabilities. 2024;**4**(4):741-753

[17] Yang W et al. Physical activity and mental health in children and adolescents with intellectual disabilities: A meta-analysis using the RE-AIM framework. The International Journal of Behavioral Nutrition and Physical Activity. 2022;**19**(1):80. DOI: 10.1186/s12966-022-01312-1

[18] Zhu G et al. The impact of physical activity on cognitive function in children and adolescents with intellectual disabilities: A meta-analysis. International Journal of Human Movement and Sports Sciences. 2023;**11**(6):1253-1265

[19] Borland RL et al. Effects of physical activity on behaviour and emotional problems, mental health and psychosocial well-being in children and adolescents with intellectual disability: A systematic review. Journal of Applied Research in Intellectual Disabilities: JARID. 2022;**35**(2):399-420. DOI: 10.1111/jar.12961

[20] Dykens EM, Rosner BA, Butterbaugh G. Exercise and sports in children and adolescents with developmental disabilities: Positive physical and psychosocial effects. Child and Adolescent Psychiatric Clinics of North America. 1998;**7**(4):757-771

[21] Scifo L et al. Sport intervention programs (SIPs) to improve health and social inclusion in people with intellectual disabilities: A systematic review. Journal of Functional Morphology and Kinesiology. 2019;**4**(3):57. DOI: 10.3390/jfmk4030057

[22] Xiao R et al. Effects of the special olympics unified sports soccer training program on executive function in adolescents with intellectual disabilities. Journal of Exercise Science and Fitness. 2024;**22**(2):103-110

[23] Chueh T-Y et al. The effects of single bouts of physical activity on cognition in adolescents and young adults with intellectual disabilities: A systematic review. Journal of Applied Research in Intellectual Disabilities. 2025;**38**(2):e70026

[24] Li Y et al. The effects of physical activity interventions in children with autism spectrum disorder: A systematic review and network meta-analysis. Review Journal of Autism and Developmental Disorders. 2023:1-15

[25] Levante A et al. The effect of sports activities on motor and social skills in autistic children and adolescents: A systematic narrative review. Current Developmental Disorders Reports. 2023;**10**(3):155-174

[26] Koh SH. Analyzing the influence of physical exercise interventions on social skills in children with autism spectrum disorder: Insights from meta-analysis. Frontiers in Psychology. 2024;**15**:1399902. DOI: 10.3389/fpsyg.2024.1399902

[27] Pasqualotto A et al. Effects of cognitive training programs on executive function in children and adolescents with autism spectrum disorder: A systematic review. Brain Sciences. 2021;**11**(10):1280

[28] Jia M et al. Benefits of exercise for children and adolescents with autism spectrum disorder: A systematic review and meta-analysis. Frontiers in Psychiatry. 2024;**15**:1462601. DOI: 10.3389/fpsyt.2024.1462601

[29] Hou Y et al. Effects of different exercise interventions on executive function in children with autism spectrum disorder: A network meta-analysis. Frontiers in Psychiatry. 2024;**15**:1440123. DOI: 10.3389/fpsyt.2024.1440123

[30] Fang Q et al. Effects of exergaming on physical and cognitive functions in individuals with autism spectrum disorder: A systematic review. Games for Health Journal. 2019;**8**(2):74-84. DOI: 10.1089/g4h.2018.0032

[31] Suárez-Manzano S et al. Effects of physical activity on cognition, behavior, and motor skills in youth with autism Spectrum disorder: A systematic review of intervention studies. Behavioral Sciences (Basel, Switzerland). 2024;**14**(4):330. DOI: 10.3390/bs14040330

[32] Zborowska AM. The role of physical activity and sport in children and adolescents with autism spectrum disorder (ASD): A narrative review. Sports Psychiatry: Journal of Sports and Exercise Psychiatry. 2024. Advance online publication. DOI: 10.1024/2674-0052/a000075

[33] Strofylla G et al. Efficacy of 6-month adapted aerobic program on motor proficiency, BMI and cognitive abilities of students with autism spectrum disorder: Three case reports. Open Journal of Social Sciences. 2021;**9**(9):153-166

[34] Koutsouki D, Asonitou A, Charitou S. Cognitive and Motor Development. Cognitive and Motor Development. [Postgraduate Textbook]. Open Academic Editions: Kallipos; 2022. DOI: 10.57713/kallipos-142

[35] Bochun Y. Effects of Exergames executive function and physical activity in young autistic groups. Frontiers in Educational Research. 2024;**7**:11

[36] Nekar DM et al. Effects of augmented reality game-based cognitive-motor training on restricted and repetitive behaviors and executive function in patients with autism spectrum disorder. Healthcare (Basel, Switzerland). 2022;**10**(10):1981. DOI: 10.3390/healthcare10101981

[37] Strofylla G et al. Profile of social skills in students with autism spectrum disorder. Advances in Physical Education. 2021;**11**(02):195

[38] Montalva-Valenzuela F et al. Effects of physical activity, exercise and sport on executive function in young people with attention deficit hyperactivity disorder: A systematic review. European Journal of Investigation in Health, Psychology and Education. 2022;**12**(1):61-76. DOI: 10.3390/ejihpe12010006

[39] Crova C et al. Cognitively challenging physical activity benefits executive function in overweight children. Journal of Sports Sciences. 2014;**32**(3):201-211. DOI: 10.1080/02640414.2013.828849

[40] Liu S et al. Effects of acute and chronic exercises on executive function in children and adolescents: A systemic review and meta-analysis. Frontiers in Psychology. 2020;**11**:554915. DOI: 10.3389/fpsyg.2020.554915

[41] Ishihara T et al. Relationship between sports experience and executive function in 6-12-year-old children: Independence from physical fitness and

moderation by gender. Developmental Science. 2018;**21**(3):e12555. DOI: 10.1111/desc.12555

[42] Freire K et al. Social exercise interventions for children who have complex developmental needs: A systematic review. Journal of Child Health Care: For Professionals Working with Children in the Hospital and Community. 2025;**29**(1):245-287. DOI: 10.1177/13674935231190984

[43] Fong SSM et al. Taekwondo training improves sensory organization and balance control in children with developmental coordination disorder: A randomized controlled trial. Research in Developmental Disabilities. 2012;**33**(1):85-95. DOI: 10.1016/j.ridd.2011.08.023

[44] Pan C-Y et al. Effects of physical activity intervention on motor proficiency and physical fitness in children with ADHD: An exploratory study. Journal of Attention Disorders. 2017;**21**(9):783-795. DOI: 10.1177/1087054714533192

[45] Smith AL et al. Pilot physical activity intervention reduces severity of ADHD symptoms in young children. Journal of Attention Disorders. 2013;**17**(1):70-82. DOI: 10.1177/1087054711417395

[46] Chou C-C, Huang C-J. Effects of an 8-week yoga program on sustained attention and discrimination function in children with attention deficit hyperactivity disorder. PeerJ. 2017;**5**:e2883. DOI: 10.7717/peerj.2883

[47] Kambas A, Venetsanou F, Kelaraki D, Karageorgopoulou M. Group psychomotor therapy improves socio-emotional and motor competence of pre-school aged children, with and without attention deficit hyperactivity disorder. Body, Movement and Dance in Psychotherapy. 2025:1-17

[48] Gearing RE et al. The effectiveness of booster sessions in CBT treatment for child and adolescent mood and anxiety disorders. Depression and Anxiety. 2013;**30**(9):800-808. DOI: 10.1002/da.22118

[49] Marwaha P, Homavazir M, Mathad P, Sharma N, Varma P. Investigation on the impact of swimming on children with ADHD. Multidisciplinary Science Journal. 2024;**6**:2024ss0618. DOI: 10.31893/multiscience.2024ss0618

[50] Yang G et al. Effect of aerobic exercise on the improvement of executive function in children with attention deficit hyperactivity disorder: A systematic review and meta-analysis. Frontiers in Psychology. 2024;**15**:1376354. DOI: 10.3389/fpsyg.2024.1376354

[51] Riper V, Stephanie M, et al. Aerobic exercise, cognitive performance, and brain activity in adolescents with attention-deficit/hyperactivity disorder. Medicine and Science in Sports and Exercise. 2023;**55**(8):1445-1455. DOI: 10.1249/MSS.0000000000003159

[52] Kadri A et al. Effect of taekwondo practice on cognitive function in adolescents with attention deficit hyperactivity disorder. International Journal of Environmental Research and Public Health. 2019;**16**(2):204. DOI: 10.3390/ijerph16020204

[53] Chimiklis AL et al. Yoga, mindfulness, and meditation interventions for youth with ADHD: Systematic review and meta-analysis. Journal of Child and Family Studies. 2018;**27**:3155-3168

[54] Cohen SCL, Harvey DJ, Shields RH, Shields GS, Rashedi RN, Tancredi DJ, et al. Effects of yoga on attention, impulsivity, and hyperactivity in preschool-aged children with

attention-deficit hyperactivity disorder symptoms. Journal of Developmental & Behavioral Pediatrics. 2018;**39**(3):200-209. DOI: 10.1097/DBP.0000000000000552

[55] Blank R et al. International clinical practice recommendations on the definition, diagnosis, assessment, intervention, and psychosocial aspects of developmental coordination disorder. Developmental Medicine and Child Neurology. 2019;**61**(3):242-285. DOI: 10.1111/dmcn.14132

[56] Meachon EJ. Perspective: Acknowledging complexity to advance the understanding of developmental coordination disorder. Frontiers in Human Neuroscience. 2023;**16**:1082209. DOI: 10.3389/fnhum.2022.1082209

[57] Asonitou K, Koutsouki D. PASS theory and movement disorders: Methodology for assessment and intervention. Children (Basel, Switzerland). 2024;**11**(10):1192. DOI: 10.3390/children11101192

[58] Asonitou K et al. Motor and cognitive performance differences between children with and without developmental coordination disorder (DCD). Research in Developmental Disabilities. 2012;**33**(4):996-1005. DOI: 10.1016/j.ridd.2012.01.008

[59] Asonitou K, Koutsouki D. Cognitive process-based subtypes of developmental coordination disorder (DCD). Human Movement Science. 2016;**47**:121-134. DOI: 10.1016/j.humov.2016.01.002

[60] Lachambre C et al. Attentional and executive functions in children and adolescents with developmental coordination disorder and the influence of comorbid disorders: A systematic review of the literature. PLoS One. 2021;**16**(6):e0252043. DOI: 10.1371/journal.pone.0252043

[61] Fujisawa S, Saito A, Sugawara M, Nakai A. Association between developmental coordination disorder traits, autistic traits, and emotional/behavioral problems in Japanese preschool children. Children. 2025;**12**(4):420. DOI: 10.3390/children12040420

[62] Lust JM et al. The subtypes of developmental coordination disorder. Developmental Medicine and Child Neurology. 2022;**64**(11):1366-1374. DOI: 10.1111/dmcn.15260

[63] Van Dyck D et al. Cognitive, perceptual, and motor profiles of school-aged children with developmental coordination disorder. Frontiers in Psychology. 2022;**13**:860766. DOI: 10.3389/fpsyg.2022.860766

[64] Burrell T. Five Different Physical Exercises That Affect the Brain in Very Different Ways. 2017. Available from: https://www.printfriendly.com/p/g/pqrhg8

[65] Papadopoulos TC, Parrila RK, Kirby JR. Cognition, Intelligence, and Achievement: A Tribute to J. P. Das. Elsevier Inc; 2015. pp. 443-458. DOI: 10.1016/B978-0-12-410388-7.00021-X. ISBN 978-0-12-410388-7

[66] Hillman CH et al. Effects of the FITKids randomized controlled trial on executive control and brain function. Pediatrics. 2014;**134**(4):e1063-e1071. DOI: 10.1542/peds.2013-3219

[67] Monastiridi SG, Katartzi ES, Kontou MG, Kourtessis T, Vlachopoulos SP. Positive relations of physical fitness and exercise intervention programs with motor competence and health-related quality of life in

developmental coordination disorder: A systematic review. European Journal of Physical Education and Sport Science [Online]. 2020. Web. 14 May 2025

[68] Tan JL et al. Impact of a multimodal exercise program on tibial bone health in adolescents with development coordination disorder: An examination of feasibility and potential efficacy. Journal of Musculoskeletal and Neuronal Interactions. 2020;**20**(4):445-471

[69] Engel AC et al. Exploring the relationship between fundamental motor skill interventions and physical activity levels in children: A systematic review and meta-analysis. Sports Medicine (Auckland, N.Z.). 2018;**48**(8):1845-1857. DOI: 10.1007/s40279-018-0923-3

[70] Zaragas H et al. The effects of physical activity in children and adolescents with developmental coordination disorder. Neurology International. 2023;**15**(3):804-820. DOI: 10.3390/neurolint15030051

[71] Smits-Engelsman B et al. Evaluating the evidence for motor-based interventions in developmental coordination disorder: A systematic review and meta-analysis. Research in Developmental Disabilities. 2018;**74**:72-102. DOI: 10.1016/j.ridd.2018.01.002

[72] Lucas BR et al. Interventions to improve gross motor performance in children with neurodevelopmental disorders: A meta-analysis. BMC Pediatrics. 2016;**16**(1):193. DOI: 10.1186/s12887-016-0731-6

[73] Cairney J, Rigoli D, Piek J. Developmental coordination disorder and internalizing problems in children: The environmental stress hypothesis elaborated. Developmental Review. 2013;**33**:224-238

[74] Mancini VO et al. The elaborated environmental stress hypothesis as a framework for understanding the association between motor skills and internalizing problems: A mini-review. Frontiers in Psychology. 2016;**7**:239. DOI: 10.3389/fpsyg.2016.00239

[75] Rezaei S, Arabameri E, Sohrabi M. Examination of the impact of an eight-week exclusive exercise on the balance of children with developmental coordination disorders. Scientific Journal of Rehabilitation Medicine. 2016;**5**:57-64

[76] Harrowell I et al. Mental health outcomes of developmental coordination disorder in late adolescence. Developmental Medicine and Child Neurology. 2017;**59**(9):973-979. DOI: 10.1111/dmcn.13469

[77] Smits-Engelsman B et al. Effects of graded exergames on fitness performance in elementary school children with developmental coordination disorder. Frontiers in Sports and Active Living. 2021;**3**:653851. DOI: 10.3389/fspor.2021.653851

[78] Pergantis P. A new era of ICTs for combating symptoms of neurodevelopmental disorders. World Journal of Biology Pharmacy and Health Sciences. 2024;**18**(1):036047. DOI: 10.30574/wjbphs.2024.18.1.0145

[79] Păsărelu CR et al. Attention-deficit/hyperactivity disorder mobile apps: A systematic review. International Journal of Medical Informatics. 2020;**138**:104133. DOI: 10.1016/j.ijmedinf.2020.104133

[80] Pergantis P. Developmental coordination disorder and the role of new technologies as intervention tool. World Journal of Advanced Research and Reviews. 2023;**19**(1):519-528

[81] Coutinho F, Bosisio ME, Brown E, Rishikof S, Skaf E, Zhang X,

et al. Effectiveness of iPad apps on visual-motor skills among children with special needs between 4y0m-7y11m. Disability and Rehabilitation. Assistive Technology. 2017;**12**(4):402-410. DOI: 10.1080/17483107.2016.1185648

[82] Purpura G et al. Use of virtual reality in school-aged children with developmental coordination disorder: A novel approach. Sensors (Basel, Switzerland). 2024;**24**(17):5578. DOI: 10.3390/s24175578

[83] Alharbi M et al. Evaluating the impact of virtual reality game training on upper limb motor performance in children and adolescents with developmental coordination disorder: A scoping review using the ICF framework. Journal of Neuroengineering and Rehabilitation. 2024;**21**(1):95. DOI: 10.1186/s12984-024-01393-y

[84] Kamberidou I, Bonias A, Patsantaras N. Sport as a means of inclusion and integration for "those of us with disabilities". European Journal of Physical Education and Sport Science. 2019. Web. 14 May 2025

Chapter 6

Multicomponent Exercise as a Health Strategy for Aging: Changes in Body Composition and BMI in Older Women

Silvia Carolina Medrano-Mena, José Raúl Hoyos-Flores, Ricardo López-García, Rosa María Cruz-Castruita and Norma Angélica Borbón-Castro

Abstract

This chapter presents the effects of multicomponent training on Body Mass Index (BMI) and body composition (fat mass, abdominal fat, and lean mass) in older women and establishes the relationship between these variables. Thirty-four women in two random groups participated: an experimental group (EG) (n = 15) that received the training program and a control group (CG) (n = 19) that did not receive training. Both groups underwent morphological measurements: anthropometry (body weight, height, and BMI) and DEXA (fat mass, lean mass, and abdominal fat) before and after the program, starting measurements at week 0 and ending at 6 months (26 weeks). The program consisted of working on conditional and coordinating abilities. Significant differences were found in the CG with a decrease in lean mass (p = .048). A decrease in BMI was found in the EG and an increase in the CG (p = .027). A low association of abdominal fat with BMI was found in the EG, and a high association of fat mass with BMI was found in the CG. Multicomponent training favors the reduction of body composition parameters related to suffering illnesses, while physical inactivity contributes to decreased lean mass, bringing consequences such as sarcopenia and sarcopenic obesity and, thus, fragility.

Keywords: body composition, BMI, DEXA, multicomponent training, older adult

1. Introduction

The increase in Body Mass Index (BMI) is a worldwide problem, with its prevalence increasing among all ages [1]. This problem is reflected in increased body fat frequently due to poor nutrition, sedentariness, and genetics [2], causing some non-communicable diseases and even death.

Physiological changes, such as body and organism changes, usually occur during aging. The most notable of these changes is a loss of muscle and bone mass and increased body fat [3]. These changes affect organism adaptation [4, 5], leading to non-communicable diseases such as diabetes, hypertension, etc., that can sometimes affect body composition and increase BMI, impairing adult mobility [3, 6].

Muscle mass decreases from age 40 to 70 or 75 years, mainly in sedentary individuals, and is substituted by body fat [7]. Thus, fat can be the main factor for muscle mass loss at these ages due to metabolic consequences. In the case of BMI, an increase in life expectancy in the Mexican and American populations is related to a BMI range of 23–30 [8]. A BMI below these values may be related to dementia, and a BMI above 30 may be related to osteoarthritis [9, 10].

Physical activity in older adults may provide health benefits if it focuses on developing different physical abilities such as strength, flexibility, resistance, and balance [11]. The opposite occurs in older adults if they are inactive. Muscle strength, endurance, mobility, and walking decrease, which increases body fat [12]. Some studies have shown that the energy balance of the diet through exercise can benefit body fat reduction and muscle mass by allowing muscle to be used as energy [13]. Other studies have shown that physical activity can produce positive changes in muscle mass and reduce body fat [14]. Therefore, physical activity in older adults can be a positive answer for health, body composition, and healthy aging, reducing the risk of morbidity and mortality [4]. This chapter aims to show the effects of a multicomponent training program on BMI and body composition (fat mass, abdominal fat, and lean mass) in older women and establish the relationship between these variables. In addition, we compare the results found in BMI and body composition with other research.

2. Biological aging

"Aging can be seen as an inevitable inherent biological program by which the maximal lifespan of each species of the animal kingdom is time restricted and to a large extent predetermined" [15]. It is difficult to determine the biological start of aging. Some authors state that aging begins in puberty and is a continuous process throughout adulthood. Others mention that this occurs from the beginning of conception. Socially, the characteristics of members of society who are perceived as old vary with cultural environments and from generation to generation [16].

The aging process is characterized by a progressive loss at the molecular, cellular, tissue, and organ levels [15]. This loss results in system alterations when the organs and tissues experience decreased activity, producing modifications and physiological alterations such as the loss of nerve cells, the loss of elasticity in tissues and blood vessels, etc., leading to an increasingly marked decrease in their normal functioning [17].

López-Otin et al. [18] proposed nine cellular and molecular traits of aging. These traits are genomic instability, telomere attrition, epigenetic alterations, loss of proteostasis, deregulated nutrient sensing, mitochondrial dysfunction, cellular senescence, stem cell exhaustion, and altered intercellular communication. In the search to define the distinctive characteristics of aging, they can contribute to and facilitate the creation of a framework for future studies on how the molecular mechanisms of aging occur and, in turn, help design interventions to improve human health and promote healthy aging. However, numerous challenges generate the need to continue research to understand the complexity involved in the biological process of aging [15, 19, 20].

2.1 Physiological changes of aging, physical inactivity, loss of muscle mass

During the life cycle, each living organism ages and concludes in death. The rate of aging varies due to the diversity of risk profiles in each individual's age-related diseases, where the idea that biological and chronological age are not entirely similar is revived [21]. Studies such as Tan et al. [22] and Mangino [23] refer to the relative contribution of genetic and non-genetic factors in monozygotic twins. On the other hand, Brooks-Wilson [24] explains that the inheritance of longevity is approximately 25%; thus, environmental factors and lifestyle play an important role. Thus, the rate of aging varies due to the interaction of genetic factors, lifestyles, and environmental factors [15].

The aging process produces physiological changes in the organism at the structural level, making the person prone to physiological consequences that put health at risk. **Table 1** presents the effects of the aging process on physiological and body structure factors, according to Landinez et al. [4].

Structural changes	Physiological consequences
Body composition: • Decrease in body water. • Decrease in organ size. • Relative increase in body fat.	• Decreased resistance to dehydration. • Altered drug distribution.
Integumentary system: • Decreased epidermal cell turnover. • Dermoepidermal and subcutaneous atrophy.	• Skin wrinkles and laxity. • Capillary fragility. • Telangiectasia.
• Decreased number of melanocytes. • Atrophy of hair follicles and sweat glands. • Decreased sebaceous gland activity. • Decreased dermal vasculature and capillary loops.	• Susceptibility to pressure ulcers. • Xerosis cutis. • Actinic keratosis. • Graying and alopecia.
Cardiovascular system: • Decreased number of myocardial cells and contractility. • Increased resistance to ventricular filling. • Decreased AV pacemaker activity, conduction velocity, and baroreceptor sensitivity.	• Decreased heart reserve. • Poor pulse response to exercise. • Arrhythmias. • Increased differential pulse pressure.
• Arterial stiffness. • Decreased cardiac output and blood flow to most organs.	• Increased arterial pressure. • Inadequate response to orthostatism. • Postural syncope.
Respiratory system: • Decreased chest and lung wall compliance, cough strength, and mucociliary clearance. • Loss of alveolar septum. • Airway collapse and increased closure volume.	• Decreased vital capacity. • Increased residual volume and alveolar-arterial oxygen differential. • Increased risk of infections and aspiration.
Renal apparatus: • Absolute decrease in the number of nephrons and kidney weight. • Decreased bladder and sphincter tone.	Reduced glomerular filtration (creatinine level remains stable due to decreased production).

Structural changes	Physiological consequences
Senses: (a) Vision: • Altered vitreous and retinal physiology. • Macular degeneration. Disorder of lens color, stiffness, and size.	• Decreased visual acuity, visual fields, and speed of adaptation to darkness. • Accommodation and pupillary reflex disorders and development of cataracts, astigmatism, and myopia.
Senses (b) Hearing: • Decreased sensory cell function in the vestibular apparatus.	• Reduced hearing (high frequencies) and sound discrimination; balance disorders.
(c) Taste and smell: • Decreased number and function of taste buds and olfactory sensory cells. • Reduced saliva production.	• Decreased gastronomic satisfaction.
Locomotor system: (a) Height: • Shortening of the spine due to disc narrowing. • Kyphosis.	• Progressive height loss. • Osteoporosis. • Vertebral collapse and long bone fracture with minimal trauma.
(b) Bones: • Universal loss of bone mass. (c) Joints: • Reduced joint elasticity. • Fibrillar degeneration of joint cartilage with atrophy and denudation of the surface. (d) Muscles: • Decreased number of muscle cells and increased muscle fat content.	

Source: Ref. [4].

Table 1.
Structural changes and physiological consequences of the aging process.

Muscle mass is an important aspect of the physiological changes during aging. This aspect is considered extremely important for the individual's general health and the many functions that it influences. In addition to providing strength, it is essential for movement, metabolism, glycogen storage, temperature regulation, joint stability, and endocrine function [25].

Functional limitations and the loss of physical activity in older adults are associated with loss of muscle strength and a decreased basal metabolic rate when there is sarcopenia. This loss causes decreased energy in the adult's body. When the energy requirement does not correspond to the adult's energy value, it progressively leads to weight gain or loss and, therefore, to the suffering and development of chronic diseases [4]. Likewise, the loss increases the risk of suffering cancer, cardiovascular diseases, neurodegenerative diseases, and diabetes, a situation that increases as age increases [15].

The modifications of soft tissues (joints, tendons, ligaments, cartilage, etc.), sedentariness, and the aging process have been considered important factors for the reduction in flexibility [26]. Range of motion loss due to sedentary behaviors causes muscle shortening due to joint inactivity. Soft tissues physiologically undergo a mechanical and biological change that promotes the development of muscle and tendon stiffness or hardness, which impedes mobility [27].

2.2 Relationship between body mass index and function

Parameters, such as weight, height, and BMI, are anthropometric measures obtained invasively as a fundamental part of obtaining general data from physical examination. Some studies state that these parameters are associated with health, disability, and mortality in older adults [28–31]. In addition, studies such as Tsai and Chang [30] and Tsai et al. [32] report that anthropometric parameters such as the mid-arm circumference and calf circumference are associated with mortality in institutionalized older adults.

During the aging process, changes in body composition occur that affect organism autonomy and function [33]; however, there is little information that provides a more specific overview that allows a more objective corroboration and verification that parameters such as BMI and arm and calf circumference are associated with loss of function in old age [34].

Other studies conducted in developed countries such as the United States and the United Kingdom mentioned that a high BMI increases the risk that an older adult will be functionally disabled. This problem increases the risk of suffering and developing chronic diseases, increasing joint wear and tear, and decreasing physical activity [28, 29, 35].

It is estimated that adults over 65 with a high BMI initially have good functional capacity but a high risk of a marked decline. This situation leads to the progressive loss of the ability to perform activities of daily living (ADL) over 4–8 years, not to mention that the calf and arm parameters are significant predictors of function, indicating that a high muscle mass and skeletal muscle strength reflect good functional performance [34]. On the other hand, Tyrovolas et al. [7] state that, starting at the age of 40, muscle mass decreases, a fact that is more relevant in sedentary individuals because an increase in fat mass replaces this reduction. This change normally increases until 70–75, and then there is a decline.

It has been shown that the optimal BMI scores that can determine an increase in life expectancy in older populations are between 23 and 30 in the Mexican and American populations [8], coinciding with the Japanese population [36], given that a BMI less than 23 is a risk factor for disability due to dementia and a BMI above 29 is a risk factor for disability due to joint disease. Two systematic reviews mention that a high BMI (≥ 29) is associated with the development of osteoarthritis [9, 10].

2.3 Multicomponent training program for health during aging

The positive effects of physical exercise on an older adult's health can be more efficiently achieved when the intervention includes more than one fitness component—in other words, strength, resistance, flexibility, or balance—in comparison to exclusive training of only one component [11]. The simultaneous work of various components is known as a multicomponent program that allows autonomously developing daily activities such as bathing, going to the bathroom, climbing stairs, getting up from a chair, walking, or using a wheelchair, allowing the adult an opportunity to have a better quality of life and lifestyle [37].

These programs have demonstrated positive changes in functional capacity by combining strength, endurance, balance, and gait training. The benefits of physical exercise are associated with a decreased mortality risk, chronic diseases, institutionalization, and cognitive and functional deterioration. Physical exercise is the most effective intervention that delays disability and functional loss [38]. Studies such as Taguchi et al. [39], whose objective was to see the effects of a 12-month multicomponent training program of physical performance and daily physical activity among very old individuals with minimal disabilities, showed how lower body strength increased and agility time decreased.

Although recent evidence shows that a multicomponent program has greater weight than traditional programs regarding the benefits it provides to the physical capacity of older adults, other studies show that interventions with both programs in the short term can effectively improve strength and agility [40]. However, the benefit that can be obtained from applying multicomponent programs is favorable even for those adults who are already fragile [41].

Just as the increase and benefit to fundamental abilities such as strength and agility are well known, Justine et al. [37], in a study with a quasi-experimental design to know the effects of the application of a multicomponent program, were able to observe and record significant increases in cardiorespiratory resistance and upper limb strength (41.79%), lower limb strength (46.19%), and balance (49.58%). Without leaving aside flexibility, they also showed improvement (right lower limb, 63.57%) in the left lower limb (44.17%), the right upper limb (36.67%), and the left upper limb (63.1%), confirming what Bouchard and Shepard [42] had mentioned in the Toronto model, which refers to the fact that the fundamental abilities that must be developed are joint flexibility, muscular strength, cardiorespiratory endurance, agility, coordination, and balance to achieve optimal functionality.

It is important to note that implementing this program in the older population is currently in short supply, even though its contributions and benefits have already been shown. Bouaziz et al. [43] support this statement by showing a systematic review from a search in CINAHL Plus, Embase, Medline, PubMed Central, ScienceDirect, Scopus, Sport Discus, and Web of Science. They considered intervention studies and obtained 2525 articles, but only 27 were included since these met the requirements of a multicomponent program.

Multicomponent program interventions are not exclusive to the healthy older adult population. They also aim to prevent the transition to frailty since preventing and delaying this process is beneficial in reducing adverse health outcomes, such as fall rates, hospitalization, institutionalization, and the associated financial costs. Studies, such as Fairhall et al. [44], have demonstrated the effectiveness of multicomponent programs in older adults with pre-frailty and diagnosed frailty and shown that multicomponent programs are profitable for this population.

3. Data collection methodology to know the changes in body composition and BMI of older women with multicomponent training

This quasi-experimental correlational explanatory study assessed a sample of 34 older women randomly divided into two groups: an EG (n = 15) (69.33 ± 4.85 years) that received treatment (multicomponent training program) and a CG (n = 19) (73.58 ± 7.68 years) that did not receive treatment. Both groups underwent morphological measurements using anthropometry and DEXA before (initial measures) and after (final measures) the training program, starting the measurements at week 0 (month 1) and the final measurement at 6 months (26 weeks). The inclusion criteria were women aged 60 years or older with the time and willingness to attend a physical training program, attending 80% of the program, and having a good clinical history with one or more controlled chronic diseases without complications. Both groups were given a consent letter informing them of the study protocol and their right to participate in the project.

Personnel certified by the International Society for Advances in Kinanthropometry (ISAK) made morphological measurements using anthropometry before and after the training program. Measurements were made at exactly 8:00 am; the older women showed up wearing appropriate clothing after fasting for at least 4 hours. Body weight in kilograms was recorded with a mechanical column scale, and height with a SECA stadiometer (model 220) according to ISAK standards.

A General Electric DEXA (GE Healthcare Lunar Technology Bone Densitometer with enCore model LU43616ES software) was used. Body weight, height, age, and gender data were captured with the software to begin the total body scan. Total body components of body fat, lean mass, and abdominal fat in kilograms, in addition to body weight and BMI, were measured. The ranges established by the World Health Organization (kg/m^2), which classifies BMI < 18.5 as underweight, 18.5–24.9 as normal weight, 25–29.9 as overweight, and > 30 as obese, were used.

Collected data were analyzed using the SPSS statistical program (IBM SPSS Statistics, Version 25, IBM Corp., Armonk, NY). The Kolmogorov-Smirnov test was used to determine the normal distribution of the variables. Descriptive statistics were presented as means and standard deviations. Basic measurements (body weight and BMI) and body composition (fat mass, lean mass, and abdominal fat) comparisons of the initial and final measurements were analyzed using Student's *t*-test for related samples. Student's t-test of an independent sample was used to compare the differences in the basic measurements and body composition obtained from the initial and final measurements between the experimental and control groups, obtaining a significant p-value ≤ .05. The Pearson correlation coefficient was used to determine the relationship between body composition variables (fat mass, lean mass, and abdominal fat) and BMI, using significance levels of $p \leq .05$, $p \leq .01$, and $p \leq .001$).

3.1 Multicomponent training program

The physical training sessions applied to the EG were conducted in the Health and Wellness Center of a public university on Monday, Wednesday, and Friday, with one-hour sessions in the morning. The multicomponent program was based on the capacity periodization system by Lanier [45] and the recommendations for physical activity for older adults established by organizations such as the American College of Sports Medicine (ACSM). Twenty-four six-month microcycles were structured and distributed in conditional (strength, resistance, and flexibility) and coordinative (coordination and balance) physical abilities.

The physical abilities of the training sessions were divided as follows: endurance was trained three times a week for 20 minutes each session with dance exercises, rhythm changes, and functional exercises; strength was trained twice a week for 20 minutes per session with exercises sitting, lying down, and standing implemented using resistance bands and self-loads; and flexibility training was performed three times a week for 20 minutes per session. The exercises were passive static and active dynamic stretching; finally, balance training was performed twice a week for 10 minutes per session. Teaching materials, such as resistance bands, balls, wooden sticks, small bags, hoops, mats, chairs, scarves, ropes, and music, were used to train abilities.

As part of the strategies to encourage adherence to the physical training program, several factors were considered, mainly those identified as facilitators for the practice of physical activity proposed by de Lacy-Vawdon et al. [46]. Therefore, in addition to considering the periodicity of attendance and the quantified follow-up of the evolution in terms of physical changes produced by the program, factors such as place (accessibility), schedule (availability), type of material (adapted to the characteristics of the target population), environment and structure of the program, tastes, preferences, and needs of the participants were taken into account, all accompanied by the follow-up and guidance of professionals in the area previously trained to provide adequate care to this population group.

4. Characteristics of the older women adults before and after the training program

4.1 BMI and body composition changes in older women

Baseline measurements of the two groups were evaluated before and after the training program (**Table 2**). Body weight (p = .290) and BMI (p = .230) decreased in the experimental group. Body weight decreased (p = .292), and BMI increased in the control group (p = .837). Regarding BMI ranges, both groups had overweight problems in the initial and final measurements.

Table 3 shows the body composition values of both groups before and after the training program. A decrease in fat mass (p = .228) and abdominal fat (p = .224) and an increase in lean mass (p = .060) were found in the experimental group. Changes in fat mass (p = .905), lean mass (p = .048), and abdominal fat (p = .516) were found in the control group.

In the comparison results between both groups regarding the differences obtained from the initial and final basic and body composition measurements, the following data were found: the experimental group obtained a greater decrease in body weight (p = .616), fat mass (p = .098), and abdominal fat (p = .860) than the control group.

Basic measurements	Initial measurement	Final measurement	p-value
EG (n = 15)			
Age (years)	69.33 ± 4.85	—	—
Height (m)	1.546 ± 0.077	—	—
Body weight (kg)	67.273 ± 12.148	66.560 ± 11.599	.299
BMI (kg/m^2)	28.54 ± 4.71	28.20 ± 4.09	.230
CG (n = 19)			
Age (years)	73.58 ± 7.68	—	—
Height (m)	1.523 ± 0.792	—	—
Body weight (kg)	66.543 ± 10.877	66.010 ± 11.356	.292
BMI (kg/m^2)	28.631 ± 4.225	28.66 ± 4.55	.837

Note. n: number; EG: experimental group; CG: control group; m: meters; kg: kilograms; BMI: body mass index; kg/m^2: kilograms by height in meters squared; $^*p \leq .05$.

Table 2.
Student t-test analysis of the basic measurements.

Body composition	Initial measurement	Final measurement	*p*-value
EG (n = 15)			
Fat mass (kg)	30.286 ± 8.555	29.476 ± 7.294	.228
Lean mass (kg)	34.692 ± 4.714	35.295 ± 4.647	.060
Abdominal fat (kg)	1.396 ± 0.510	1.335 ± 0.437	.224
CG (n = 19)			
Fat mass (kg)	29.162 ± 7.014	29.120 ± 7.614	.905
Lean mass (kg)	35.566 ± 5.687	35.074 ± 5.604	.048*
Abdominal fat (kg)	1.279 ± 0.491	1.243 ± 0.469	.516

*Note. N: number; EG: experimental group; CG: control group; kg: kilogram; * $p \leq .05$.*

Table 3.
Student's t-test analysis of body composition.

In the experimental group, the BMI decreased, and in the control group, it increased with a significant difference (p = .027). Lean mass in the experimental group increased, while in the control group, it decreased (p = .860) (**Table 4**).

4.2 Relationship between BMI and body composition in older women

Pearson's correlation analysis (**Table 5**) was performed, relating body composition to the BMI of both groups' initial and final measurements. A very high association of fat mass with BMI in the initial (r = .930; $p \leq .001$) and final measurements (r = .922; $p \leq .001$), and a high association of fat mass with BMI in the initial (r = .780; $p \leq .01$) and final measurements (r = .710; $p \leq .01$) were found in the experimental group. A moderate association was found between abdominal fat and BMI in the initial measurement (r = .560; $p \leq .05$), and a low association in the final measurement (r = .395; p = .145).

A high association between fat mass and BMI at the initial measurement (r = .798; $p \leq .001$) and a very high association at the final measurement (r = .809; $p \leq .001$) were found in the control group. Fat mass had a moderate association with BMI at the initial measurement (r = .523; $p \leq .05$) and final measurement (r = .533; $p \leq .05$) and a high association between abdominal fat and BMI at the initial (r = .778; $p \leq .001$) and final measurement (r = .734; $p \leq .001$).

Basic measurements and body composition	EG (n = 15)	CG (n = 19)	*p*-value
Body weight (kg)	−0.713 ± 2.561	−0.532 ± 2.141	.616
BMI (kg/m^2)	−0.346 ± 1.068	0.031 ± 0.658	.027*
Fat mass (kg)	−0.810 ± 2.488	−0.042 ± 1.523	.098
Lean mass (kg)	0.603 ± 1.144	−0.492 ± 1.012	.455
Abdominal fat (kg)	−0.061 ± 0.186	−0.035 ± 0.234	.860

*Note. n: number; EG: experimental group; CG: control group; kg: kilograms; BMI: body mass index; kg/m^2: kilograms by height in meters squared; * $p \leq .05$.*

Table 4.
Comparison of the differences between the EG and the CG from the initial to the final measurement of basic measurements and body composition.

Body composition		**Initial measurement**	**Final measurement**
		BMI (kg/m^2)	
EG (n = 15)			
Fat mass (kg)	*r*	.930***	.922***
Lean mass (kg)	*r*	.780**	.710**
Abdominal fat (kg)	*r*	.560*	.395
CG (n = 19)			
Fat mass (kg)	*r*	.798***	.809***
Lean mass (kg)	*r*	.523*	.533*
Abdominal fat (kg)	*r*	.778***	.734***

*Note. n: number; EG: experimental group; CG: control group; kg: kilograms; BMI: body mass index; kg/m²: kilograms by height in meters squared; Pearson coefficient; * p ≤ .05; ** p ≤ .01; *** p ≤ .001.*

Table 5.
Pearson correlation analysis between body composition and BMI of both measurements.

4.3 Comparison of changes found in BMI and body composition with other research

The results of this study show that after 6 months of training, the EG managed to reduce body weight, BMI, fat mass, and abdominal fat and obtain an increase in lean mass, similar to the control group, which did not carry out the training program; however, there was an increase in BMI and a significant decrease in lean mass in the CG.

These results are similar to Bemben and Bemben's [47] 40-week endurance training program and Maddalozzo and Snow's [48] 24-week moderate- and high-intensity training program that decreased fat mass and increased lean mass with significant changes. Other studies have found no increases in lean mass or decreases in fat mass in year-long training programs [49, 50]. These findings may be due to a high nutritional caloric intake since increases in body weight were also found.

A 6-month study [14] with a supplementation intervention of protein, vitamins, and minerals found significant changes with decreases in body weight and fat mass and increases in lean mass. Nevertheless, another study with a training and interventional nutrition program did not report significant changes in body composition [51].

Cancela et al. [52] mention the results of an 8-month training program of 60 minutes with two sessions per week. They found significant changes in body composition without a supplementation intervention. These findings could be due to the 8-month duration of the training sessions.

Other results from this study were the difference in body composition changes after 6 months of training between both groups, where the experimental group had a decrease in BMI and the control group an increase, with significant differences. These data are similar to Rondelli et al. [53], with 12 months of training divided into two groups (supplementation and placebo groups), and Kang et al. [54], with 4 weeks of training. BMI decreased in both studies, although in the supplementation group in Rondelli et al. [53], the BMI increased, but as a curious fact, their fat mass decreased. These findings may be because the supplementation included protein intake, a macronutrient of importance for growth, development, and muscle mass maintenance. Supplementation may have increased lean mass and, therefore, BMI.

Other studies show the opposite about fat mass. The research by Tieland et al. [55], with a duration of 24 weeks of training divided into two groups (protein supplementation and placebo groups), found increases in fat mass in the supplementation group and a decrease in the placebo group; however, in lean mass, the supplementation group had increases and the placebo group a decrease. This finding shows that protein supplementation is more effective in gaining muscle mass than decreasing body fat.

In the correlation analysis between body composition and BMI, our study showed significant relationships in the EG with very high associations in fat mass and high associations in lean mass with BMI in both measurements. Regarding abdominal fat, after a six-month training period, this variable does not usually have a significant relationship with BMI since low associations were found in the final measurement compared to the initial measurement, which showed a moderate association. This finding shows that after 6 months of training, abdominal fat is not usually a predominant factor in subjects with a high BMI [56].

In contrast, in the CG, moderate associations were found in lean mass and high associations in abdominal fat in both measurements with the BMI. Regarding fat mass, high associations were found in the initial measurement, and very high associations were found in the final measurement with the BMI. These data explain that after 6 months, fat mass is a very important indicator of having a high BMI in women who did not carry out a physical activity program [51, 57, 58].

There is no doubt that in the literature found, most of the studies that use nutritional supplementation as a part of a training program usually have favorable changes in body composition, such as improved lean mass, decreased fat mass, body weight, and even BMI [53, 59, 60].

5. Conclusion

Food, age, and physical activity often condition the morphological status of the subject. The changes of the EG explain that implementing a multicomponent training program in the older population favors body weight reduction, BMI, fat mass, and abdominal fat. The results of the CG and the literature reviewed show that physical inactivity can increase fat mass and decrease lean mass, bringing consequences in older adults, such as sarcopenia and sarcopenic obesity, causing complications, such as frailty.

The multicomponent training program is viable for achieving positive changes in body composition. However, conducting a study on nutrition and food supplementation is important to increase the results.

Author details

Silvia Carolina Medrano-Mena[1], José Raúl Hoyos-Flores[1], Ricardo López-García[1], Rosa María Cruz-Castruita[1*] and Norma Angélica Borbón-Castro[2]

1 Faculty of Sports Organization, Autonomous University of Nuevo Leon, San Nicolas de los Garza, Nuevo Leon, Mexico

2 Educational Program Bachelor's Degree in Sports Training, Sonora State University, Hermosillo, Sonora, Mexico

*Address all correspondence to: rosa.cruzcst@uanl.edu.mx

References

[1] Ruiz-Cota P, Bacardí-Gascón M, Jiménez-Cruz A. Historia, tendencias y causas de la obesidad en México. Revista de resultados negativos y no positivos. 2019;**4**(7):737-745

[2] Lee CH, Lee JH, Son JW, Kim U, Park JS, Lee J, et al. Normative values of short-term heart rate variability parameters in Koreans and their clinical value for the prediction of mortality. Heart, Lung and Circulation. 2018;**27**(5):576-587

[3] Ferrucci L, Hesdorffer C, Bandinelli S, Simonsick EM. Frailty as a nexus between the biology of aging, environmental conditions and clinical geriatrics. Public Health Reviews. 2010;**32**:475-488

[4] Landinez Parra NS, Contreras Valencia K, Castro Villamil Á. Proceso de envejecimiento, ejercicio y fisioterapia. Revista Cubana de Salud Pública. 2012;**38**(4):562-580

[5] Santanasto AJ, Glynn NW, Lovato LC, Blair SN, Fielding RA, Gill TM, et al. Effect of physical activity versus health education on physical function, grip strength and mobility. Journal of the American Geriatrics Society. 2017;**65**(7):1427-1433

[6] Ruggiero C, Metter EJ, Melenovsky V, Cherubini A, Najjar SS, Ble A, et al. High basal metabolic rate is a risk factor for mortality: The Baltimore longitudinal study of aging. The Journals of Gerontology Series A: Biological Sciences and Medical Sciences. 2008;**63**(7):698-706

[7] Tyrovolas S, Koyanagi A, Olaya B, Ayuso-Mateos JL, Miret M, Chatterji S, et al. The role of muscle mass and body fat on disability among older adults: A cross-national analysis. Experimental Gerontology. 2015;**69**:27-35

[8] Kumar A, Karmarkar AM, Tan A, Graham JE, Arcari CM, Ottenbacher KJ, et al. The effect of obesity on incidence of disability and mortality in Mexicans aged 50 years and older. Salud publica de Mexico. 2015;**57**:s31-s38

[9] Jiang L, Tian W, Wang Y, Rong J, Bao C, Liu Y, et al. Body mass index and susceptibility to knee osteoarthritis: A systematic review and meta-analysis. Joint, Bone, Spine. 2012;**79**(3):291-297

[10] Yusuf E, Nelissen RG, Ioan-Facsinay A, Stojanovic-Susulic V, DeGroot J, Van Osch G, et al. Association between weight or body mass index and hand osteoarthritis: A systematic review. Annals of the Rheumatic Diseases. 2010;**69**(4):761-765

[11] Cadore EL, Rodríguez-Mañas L, Sinclair A, Izquierdo M. Effects of different exercise interventions on risk of falls, gait ability, and balance in physically frail older adults: A systematic review. Rejuvenation Research. 2013;**16**(2):105-114

[12] Greve J, Alonso A, Bordini ACP, Camanho GL. Correlation between body mass index and postural balance. Clinics. 2007;**62**:717-720

[13] Visser M, Deeg DJ, Lips P. Low vitamin D and high parathyroid hormone levels as determinants of loss of muscle strength and muscle mass (sarcopenia): The longitudinal aging study Amsterdam. The Journal of Clinical Endocrinology and Metabolism. 2003;**88**(12):5766-5772

[14] Englund DA, Kirn DR, Koochek A, Zhu H, Travison TG, Reid KF, et al. Nutritional supplementation with physical activity improves muscle composition in mobility-limited older adults, the VIVE2 study: A randomized, double-blind, placebo-controlled trial. The Journals of Gerontology: Series A. 2018;**73**(1):95-101

[15] Declerck K, Berghe WV. Back to the future: Epigenetic clock plasticity towards healthy aging. Mechanisms of Ageing and Development. 2018;**174**:18-29

[16] Goswami S, Sahai M. A study of morbidity pattern in elderly population. European Researcher. 2016;**103**:119-128

[17] Velásquez M, Prieto B, Contreras RE. El envejecimiento y los radicales libres. Ciencias. 2004;**75**(43):36-43

[18] López-Otín C, Blasco MA, Partridge L, Serrano M, Kroemer G. The hallmarks of aging. Cell. 2013;**153**(6):1194-1217

[19] Martin GM. The biology of aging: 1985-2010 and beyond. The FASEB Journal. 2011;**25**(11):3756-3762

[20] Miller RA. Genes against aging. Journals of Gerontology Series A: Biomedical Sciences and Medical Sciences. 2012;**67**(5):495-502

[21] Blaha MJ, Hung RK, Dardari Z, Feldman DI, Whelton SP, Nasir K, et al. Age-dependent prognostic value of exercise capacity and derivation of fitness-associated biologic age. Heart. 2016;**102**:409-410

[22] Tan Q, Christiansen L, Thomassen M, Kruse TA, Christensen K. Twins for epigenetic studies of human aging and development. Ageing Research Reviews. 2013;**12**(1):182-187

[23] Mangino M. Genomics of ageing in twins. Proceedings of the Nutrition Society. 2014;**73**(4):526-531

[24] Brooks-Wilson AR. Genetics of healthy aging and longevity. Human Genetics. 2013;**132**(12):1323-1338

[25] Pedersen BK, Febbraio MA. Muscles, exercise and obesity: Skeletal muscle as a secretory organ. Nature Reviews Endocrinology. 2012;**8**(8):457-465

[26] Barbosa AR, Souza JM, Lebrão ML, Laurenti R, Marucci MDFN. Functional limitations of Brazilian elderly by age and gender differences: Data from SABE survey. Cadernos de Saúde Pública. 2005;**21**(4):1177-1185

[27] Correia M, Menêses A, Lima A, Cavalcante B, Ritti-Dias R. Efeito do treinamento de força na flexibilidade: Uma revisão sistemática. Revista Brasileira de Atividade Física and Saúde. 2014;**19**(1):3

[28] Lang IA, Llewellyn DJ, Alexander K, Melzer D. Obesity, physical function, and mortality in older adults. Journal of the American Geriatrics Society. 2008;**56**(8):1474-1478

[29] Al Snih S, Ottenbacher KJ, Markides KS, Kuo YF, Eschbach K, y Goodwin, J. S. The effect of obesity on disability vs mortality in older Americans. Archives of Internal Medicine. 2007;**167**(8):774-780

[30] Tsai AC, Chang TL. The effectiveness of BMI, calf circumference and mid-arm circumference in predicting subsequent mortality risk in elderly Taiwanese. British Journal of Nutrition. 2011;**105**(2):275-281

[31] Woo J, Ho SC, Sham A. Longitudinal changes in body mass index and body composition over 3 years and

relationship to health outcomes in Hong Kong Chinese age 70 and older. Journal of the American Geriatrics Society. 2001;**49**(6):737-746

[32] Tsai ACH, Lai MC, Chang TL. Mid-arm and calf circumferences (MAC and CC) are better than body mass index (BMI) in predicting health status and mortality risk in institutionalized elderly Taiwanese. Archives of Gerontology and Geriatrics. 2012;**54**(3):443-447

[33] Preto LSR, Gomes JRL, Novo AFMP, Mendes MER, Granero-Molina J. Efeitos de um programa de enfermagem de reabilitação na aptidão funcional de idosos institucionalizados. Revista de Enfermagem Referência. 2016;**8**:55

[34] Tsai HJ, Chang FK. Associations between body mass index, mid-arm circumference, calf circumference, and functional ability over time in an elderly Taiwanese population. PLoS One. 2017;**12**(4):1-11

[35] Aoki Y, Yoon SS, Chong Y, Carroll MD. Hypertension, abnormal cholesterol, and high body mass index among non-Hispanic Asian adults: United States, 2011-2012. NCHS Data Brief. 2014;**140**:1-8

[36] Zhang S, Tomata Y, Sugiyama K, Kaiho Y, Honkura K, Watanabe T, et al. Body mass index and the risk of incident functional disability in elderly Japanese: The OHSAKI cohort 2006 study. Medicine. 2016;**95**(31):1-8

[37] Justine M, Hamid TA, Mohan V, Jagannathan M. Effects of multicomponent exercise training on physical functioning among institutionalized elderly. International Scholarly Research Notices. 2012;**2012**(1):124916

[38] Izquierdo M, Cadore EL, Casas Herrero A. Ejercicio físico en el anciano frágil: Una manera eficaz de prevenir la dependencia. Kronos. 2014;**13**(1):74-81

[39] Taguchi N, Higaki Y, Inoue S, Kimura H, Tanaka K. Effects of a 12-month multicomponent exercise program on physical performance, daily physical activity, and quality of life in very elderly people with minor disabilities: An intervention study. Journal of Epidemiology. 2010;**20**(1):21-29

[40] Sousa N, Mendes R. Comparison of effects of resistance and multicomponent training on falls prevention in institutionalized elderly women. Journal of the American Geriatrics Society. 2015;**63**(2):396-397

[41] Tarazona-Santabalbina FJ, Gómez-Cabrera MC, Pérez-Ros P, Martínez-Arnau FM, Cabo H, Tsaparas K, et al. A multicomponent exercise intervention that reverses frailty and improves cognition, emotion, and social networking in the community-dwelling frail elderly: A randomized clinical trial. Journal of the American Medical Directors Association. 2016;**17**(5):426-433

[42] Bouchard C, Shepard T. Physical activity, fitness, and health: The model and key concepts. In: Bouchard C, Shephard R, Stephens T, editors. Physical Activity, Fitness and Health. Champaign, Illinois: Human Kinetics Publishers; 1993

[43] Bouaziz W, Lang PO, Schmitt E, Kaltenbach G, Geny B, Vogel T. Health benefits of multicomponent training programmes in seniors: A systematic review. International Journal of Clinical Practice. 2016;**70**(7):520-536

[44] Fairhall N, Kurrle SE, Sherrington C, Lord SR, Lockwood K, John B, et al. Effectiveness of a multifactorial intervention on preventing development

of frailty in pre-frail older people: Study protocol for a randomised controlled trial. BMJ Open. 2015;**5**(2):1-8. DOI: 10.1136/bmjopen-2014-007091

[45] Lanier A. Metodología, planeación y control del entrenamiento deportivo. 2a ed. San Antonio, México: SECCADE Imprenta; 2005. pp. 93-124

[46] de Lacy-Vawdon CJ, Klein R, Schwarzman J, Nolan G, de Silva R, Menzies D, et al. Facilitators of attendance and adherence to group-based physical activity for older adults: A literature synthesis. Journal of Aging and Physical Activity. 2018;**26**(1):155-167

[47] Bemben DA, Bemben MG. Effects of resistance exercise and body mass index on lipoprotein-lipid patterns of postmenopausal women. The Journal of Strength and Conditioning Research. 2000;**14**(1):80-85

[48] Maddalozzo GF, Snow CM. High intensity resistance training: Effects on bone in older men and women. Calcified Tissue International. 2000;**66**:399-404

[49] Going S, Lohman T, Houtkooper L, Metcalfe L, Flint-Wagner H, Blew R, et al. Effects of exercise on bone mineral density in calcium-replete postmenopausal women with and without hormone replacement therapy. Osteoporosis International. 2003;**14**:637-643

[50] Maddalozzo GF, Widrick JJ, Cardinal BJ, Winters-Stone KM, Hoffman MA, Snow CM. The effects of hormone replacement therapy and resistance training on spine bone mineral density in early postmenopausal women. Bone. 2007;**40**(5):1244-1251

[51] Fernández-Villa JM, Márquez DX, Sánchez-Garrido N, Pérez-Zepeda MU, González-Lara M. Association of healthy habits beliefs and mortality in older adults: A longitudinal analysis of the Mexican health and aging study. Journal of Aging and Health. 2017;**29**(6):973-985

[52] Cancela BL, Neves MED, Rodrigues LL, Gomes Dias AC. The influence of corporate governance on corporate sustainability: New evidence using panel data in the Iberian macroeconomic environment. International Journal of Accounting and Information Management. 2020;**28**(4):785-806

[53] Rondanelli M, Klersy C, Terracol G, Talluri J, Maugeri R, Guido D, et al. Whey protein, amino acids, and vitamin D supplementation with physical activity increases fat-free mass and strength, functionality, and quality of life and decreases inflammation in sarcopenic elderly. The American Journal of Clinical Nutrition. 2016;**103**(3):830-840

[54] Kang S, Hwang S, Klein AB, Kim SH. Multicomponent exercise for physical fitness of community-dwelling elderly women. Journal of Physical Therapy Science. 2015;**27**(3):911-915

[55] Tieland M, van de Rest O, Dirks ML, van der Zwaluw N, Mensink M, van Loon LJ, et al. Protein supplementation improves physical performance in frail elderly people: A randomized, double-blind, placebo-controlled trial. Journal of the American Medical Directors Association. 2012;**13**(8):720-726

[56] Díaz J, Espinoza-Navarro O, Pino A. Características antropométricas y fisiológicas de adultos mayores de la comuna de Arica-Chile. International Journal of Morphology. 2015;**33**(2):580-585

[57] Smith L, Tully M, Jacob L, Blackburn N, Adlakha D, Caserotti P, et al. The association between sedentary

behavior and sarcopenia among adults aged ≥ 65 years in low-and middle-income countries. International Journal of Environmental Research and Public Health. 2020;**17**(5):1708

[58] Solano-García W, Carazo-Vargas P. Intervenciones con ejercicio contra resistencia en la persona adulta mayor diagnosticada con sarcopenia. Una revisión sistemática. Pensar en Movimiento: Revista de ciencias del ejercicio y la salud. 2018;**16**(1):1-19

[59] Oh ES, Fong TG, Hshieh TT, Inouye SK. Delirium in older persons: Advances in diagnosis and treatment. JAMA. 2017;**318**(12):1161-1174

[60] Yang A, Lv Q, Chen F, Wang Y, Liu Y, Shi W, et al. The effect of vitamin D on sarcopenia depends on the level of physical activity in older adults. Journal of Cachexia, Sarcopenia and Muscle. 2020;**11**(3):678-689

Chapter 7

Perspective Chapter: Empowerment through Movement – Enhancing Psychosocial Well-Being in People with Physical Disabilities

Hafifi Hisham, Mohd Fahmi Aluwi and Nor Azlin Mohd Nordin

Abstract

This chapter explores the transformative role of movement in enhancing the psychosocial well-being of individuals with physical disabilities. Physical disabilities significantly impact mobility, independence, and quality of life, often leading to mental health challenges and social isolation. Movement emerges as a cornerstone of empowerment, offering physical, mental, and social benefits. Regular physical activity improves strength, endurance, and flexibility, while reducing secondary health complications, such as obesity and cardiovascular issues. On a psychological level, it alleviates stress, anxiety, and depression, fostering self-esteem and resilience. Socially, inclusive activities, such as adaptive sports, build camaraderie and reduce isolation. The chapter highlights barriers, including infrastructure, financial constraints, and societal stigma, and emphasizes the role of adaptive technologies and advocacy in overcoming them. Practical recommendations are provided to inspire individuals, caregivers, and healthcare professionals to view the movement as a pathway to empowerment, enabling people with physical disabilities to lead fulfilling, independent lives.

Keywords: physical disabilities, psychosocial well-being, empowerment through movement, adaptive physical activity, inclusive rehabilitation

1. Introduction

Physical disabilities affect millions of people worldwide and significantly impact their physical, psychological, and social well-being [1]. These disabilities, which can result from congenital conditions, injuries, or illnesses, often limit mobility, independence, and participation in daily activities. While medical and rehabilitative care are essential, a growing body of evidence highlights the role of movement and physical activity in empowering individuals with physical disabilities. Movement not only improves physical health, but also enhances mental resilience [2], fosters social connections, and contributes to an improved quality of life [3]. **Figure 1** shows the movement as empowerment toward quality of life, physical health, social connections, and mental resilience.

2. Types of physical disabilities

Physical disabilities impair an individual's ability to perform movements or tasks requiring physical exertion. The impact of physical disabilities varies widely, depending on the nature and severity of the condition, but they all present unique challenges that affect mobility, independence, and overall quality of life. Physical disabilities encompass a broad spectrum of conditions that impair mobility, agility, and overall physical functioning. These disabilities can arise from various causes, including congenital conditions, injuries, or chronic illnesses, and they manifest in diverse ways depending on the individual. Understanding the different types of physical disabilities is crucial for tailoring interventions, promoting inclusivity, and providing appropriate support. By categorizing these disabilities, we can better address the unique challenges faced by individuals and develop effective strategies to enhance their quality of life. The following section explores the major types of physical disabilities and their defining characteristics.

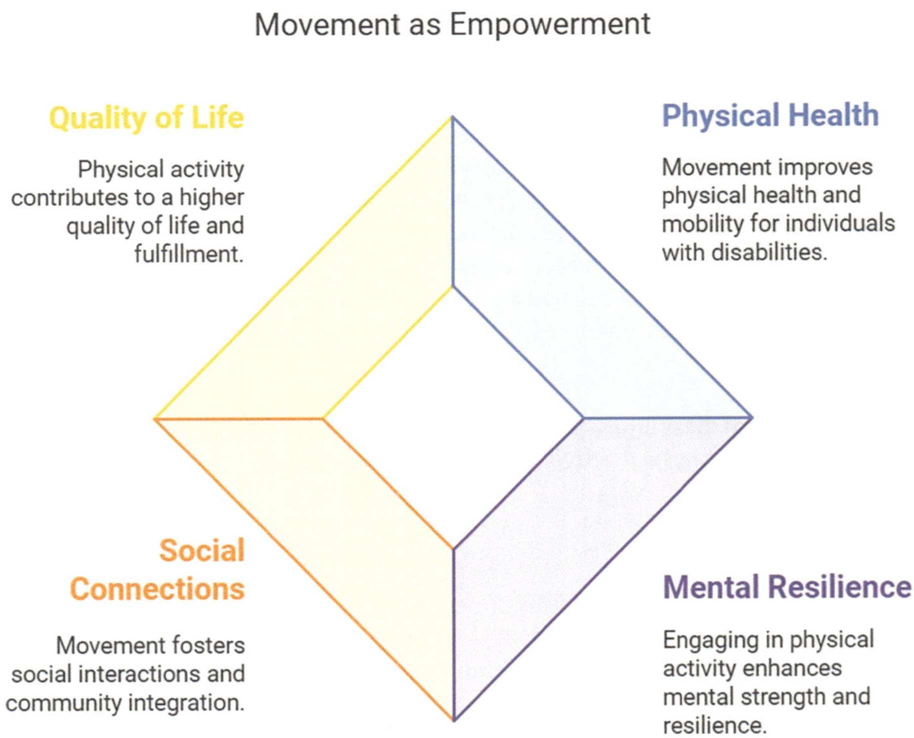

Figure 1.
The diagram of movement as empowerment on quality of life, physical health, social connections, and mental resilience among people with disabilities (PwD).

2.1 Mobility impairments

These include conditions that limit or completely hinder movement. This category includes paralysis, and conditions such as muscular dystrophy or cerebral palsy. Mobility impairments affect an individual's ability to walk, climb, or even perform routine tasks like standing or sitting. Wheelchair users or individuals who require walking aids are examples of those affected by mobility impairments.

2.2 Neurological disabilities

Neurological disabilities, such as stroke, spinal cord injuries, multiple sclerosis, and Parkinson's disease, impact the nervous system's ability to send messages to the muscles, thereby impairing movement. These disabilities often result in partial or total loss of motor function, muscle weakness, and coordination problems, significantly affecting an individual's physical independence.

2.3 Musculoskeletal disabilities

Musculoskeletal disabilities refer to conditions that affect the bones, joints, muscles, ligaments, or tendons, resulting in pain, weakness, or stiffness. Conditions such as arthritis, fractures, and osteoporosis fall under this category. These disabilities may limit an individual's ability to perform physical tasks, causing chronic pain and reduced mobility.

2.4 Amputations

Amputation refers to the loss of a limb, which can occur due to trauma, disease, or congenital conditions. Individuals with amputations face significant challenges related to mobility, balance, and performing everyday tasks, often requiring prosthetics or assistive devices for functional restoration. Each of these types of disabilities brings distinct challenges, requiring tailored strategies for care, rehabilitation, and empowerment. Understanding the different categories of physical disabilities is crucial for developing targeted interventions that promote physical, mental, and emotional well-being.

3. Prevalence of physical disabilities

Physical disabilities affect a significant proportion of the global population, with varying prevalence across different regions and demographics. These disabilities, resulting from congenital conditions, illnesses, or injuries, present challenges that impact physical functioning, independence, and quality of life. Understanding the prevalence and types of physical disabilities is essential for identifying the scale of the issue and designing targeted interventions. This section explores global and regional statistics on physical disabilities, including their prevalence in regions such as North America, Europe, Africa, Asia, and Southeast Asia. It also categorizes the primary types of physical disabilities, highlighting their characteristics and implications. Physical disabilities represent a substantial public health concern worldwide, with the World Health Organization (WHO) estimating that over 1 billion people globally live with some form of disability, and about 190 million experience significant difficulties in functioning [4]. This statistic underscores the scale of physical disabilities and the need for effective policies and interventions to address the challenges faced by affected individuals [4].

The global prevalence of physical disabilities is influenced by factors such as population aging, the rise of chronic health conditions, and advances in medical interventions that enable individuals with severe injuries or conditions to survive and live longer lives. According to WHO's Global Disability Action Plan (2014–2021), physical disabilities, including mobility impairments, are among the most common forms of disability globally. Conditions such as spinal cord injuries, amputations, cerebral palsy, and muscular dystrophy contribute significantly to the global burden of disability [5].

The prevalence of physical disabilities is often higher in low- and middle-income countries due to factors such as inadequate healthcare infrastructure [6], higher rates of injuries from road accidents [7] and occupational hazards, and untreated congenital conditions. However, developed nations also report significant prevalence rates due to aging populations and lifestyle-related conditions such as diabetes and stroke. The prevalence of physical disabilities varies significantly across different regions due to factors such as socioeconomic conditions, healthcare infrastructure, cultural attitudes, and environmental challenges. Understanding these differences is critical for tailoring interventions and allocating resources effectively.

3.1 The United States

In the Americas, an estimated 15% of the population lives with some form of disability, with physical disabilities being particularly prominent. The United States reports a prevalence of physical disabilities among 12.8% of adults [7], primarily due to conditions such as arthritis, back problems, and mobility impairments. Access to healthcare and rehabilitation services in North America is generally better than in other regions, although disparities exist among different socioeconomic groups.

3.2 Europe

Europe has a similar prevalence rate, with physical disabilities affecting approximately 80 million people, or 15% of the population [4]. Aging is a significant factor, with many disabilities arising from age-related conditions such as osteoporosis and stroke. The European Disability Strategy (2021–2030) has been pivotal in improving accessibility and integration for individuals with physical disabilities in education, employment, and healthcare.

3.3 Africa

In Africa, physical disabilities are often the result of preventable causes such as infectious diseases (e.g., polio), malnutrition, and inadequate maternal healthcare [4]. The prevalence rates vary widely across the continent, but it is estimated that up to 10% of the population has some form of disability. A lack of infrastructure and rehabilitation services poses significant challenges for individuals with physical disabilities in this region.

3.4 Asia

Asia, with its vast and diverse population, reports high numbers of individuals with physical disabilities. In countries like China and India, disabilities due to road traffic accidents and occupational hazards are highly prevalent. Cultural stigma and limited access to rehabilitation services are the main barriers that many individuals face. In East Asia, countries such as Japan are grappling with the challenges posed by the aging of the population, leading to higher prevalence rates of aged-related mobility impairments [4].

3.4.1 Southeast Asia

Southeast Asia faces unique challenges due to a combination of natural disasters, road traffic accidents, and infectious diseases like tuberculosis leading to some forms of impairments and physical disabilities [7]. Countries like Malaysia, Indonesia, and the Philippines report physical disabilities in about 10–15% of their populations [4]. Limited healthcare infrastructure including rehabilitation services and societal stigma further exacerbate the difficulties faced by individuals with disabilities in this region.

4. Cultural, regional, and socioeconomic variations in barriers to physical activity among people with physical disabilities across continents

The experiences and challenges faced by individuals with physical disabilities differ significantly across continents, influenced by cultural norms, regional contexts, and socioeconomic disparities. Understanding these variations is crucial for designing inclusive strategies that account for global diversity and promote equitable access to physical activity.

4.1 Africa

In many parts of Africa, people with physical disabilities face profound barriers due to inadequate infrastructure, limited healthcare services, and entrenched cultural stigmas. Rural areas often lack accessible transportation and public facilities, making mobility and participation in physical activities difficult [8]. Furthermore, societal perceptions of disability as a curse or misfortune exacerbate stigma, leading to exclusion from community activities [9–11]. Financial barriers are also prevalent, as poverty limits access to assistive technologies and rehabilitation services [6, 12]. However, grassroots initiatives, such as the *African Youth with Disabilities Network* (AYWDN), have started promoting disability rights and inclusive sports programs to challenge these barriers [13, 14].

4.2 Asia

In Asia, rapid urbanization and economic growth have led to both challenges and opportunities for people with physical disabilities. While modern cities like Tokyo and Singapore have made strides in accessibility, many developing countries in the region still struggle with inadequate infrastructure and social stigma [15, 16]. In rural parts of South Asia, such as India and Nepal, people with disabilities often lack access to rehabilitation services and adaptive sports programs due to geographic and financial constraints [17–20]. Cultural factors also play a role, as societal expectations may discourage individuals, particularly women, from participating in physical activities. Programs like the *Japan Para Sports Association* and *India's Paralympic Committee* are working to foster inclusivity through accessible sports initiatives [21, 22].

4.3 Europe

Europe, with its strong emphasis on human rights and social welfare, offers more supportive environments for individuals with physical disabilities. Countries like Sweden and the Netherlands are known for their inclusive urban designs, accessible public transportation, and widespread adaptive sports programs [23–25]. However, disparities exist between Western and Eastern Europe. In some Eastern European countries, limited funding and outdated infrastructure create challenges for individuals seeking physical activity opportunities. Advocacy organizations, such as the *European Disability Forum*, play a vital role in promoting accessibility and awareness, ensuring that policies support inclusivity across the continent [26].

4.4 North America

North America is recognized for its advanced adaptive sports programs and rehabilitation services, particularly in the United States and Canada [27]. However, significant disparities remain, particularly for individuals from low-income and minority communities. While urban centers may offer accessible facilities and specialized programs, rural and indigenous communities often face limited access to resources. Moreover, healthcare costs in the U.S. pose financial barriers for those without adequate insurance, limiting their access to rehabilitation and assistive technologies. Organizations such as the *National Centre on Health, Physical Activity and Disability* (NCHPAD) work to address these disparities by promoting inclusive physical activity initiatives [28].

4.5 South America

In South America, people with physical disabilities often encounter challenges stemming from socioeconomic inequality, political instability, and a lack of public awareness about disability rights. In many countries, accessible infrastructure and adaptive sports programs are limited, particularly in rural and economically disadvantaged areas. However, initiatives such as *Brazil's Paralympic Movement* have gained traction, using sports as a platform to promote inclusion and challenge societal stigma [29, 30]. Public campaigns and events, like the *Parapan American Games*, have also helped raise awareness and foster a culture of acceptance and support [31].

4.6 Oceania

Oceania, encompassing countries like Australia and New Zealand, demonstrates a strong commitment to promoting inclusivity for people with physical disabilities [32, 33]. Both countries have well-established disability support systems, including accessible infrastructure, comprehensive rehabilitation services, and adaptive sports programs. However, challenges persist in remote and indigenous communities, where geographic isolation and cultural differences limit access to services. The *National Disability Insurance Scheme (NDIS)* in Australia is a notable example of a program addressing these gaps, providing financial support, and promoting inclusive community initiatives [34].

The experiences and challenges of people with physical disabilities vary greatly across continents due to differences in cultural norms, regional infrastructure, and socioeconomic conditions. Acknowledging these global disparities allows for the development of tailored strategies that promote inclusivity and accessibility. By learning from successful initiatives worldwide and adapting them to local contexts, societies can create a more equitable environment where individuals with physical disabilities can fully participate in physical activities and lead fulfilling lives.

5. Physical health challenges

Individuals with physical disabilities commonly face unique health challenges that extend beyond their primary condition. These challenges can significantly impact their overall well-being and quality of life. Secondary health conditions and issues with independence and mobility are common, underscoring the need for proactive management and support.

5.1 Secondary health conditions

Physical disabilities often predispose individuals to secondary health conditions that further complicate their overall health and well-being. These conditions include obesity, cardiovascular problems, pressure sores, and musculoskeletal complications. **Figure 2** shows the diagram to understand the health challenges among people with disabilities (PwD).

5.1.1 Obesity

Limited mobility among individuals with physical disabilities reduces energy expenditure, increasing the risk of obesity [8]. Obesity, in turn, exacerbates mobility difficulty, leading to a vicious cycle of reduced physical activity and weight gain. According to the World Health Organization (WHO), obesity prevalence among

Figure 2.
The diagram to understand the health challenges among people with disabilities (PwD).

individuals with physical disabilities is significantly higher than in the general population, posing risks of diabetes, hypertension, and metabolic syndrome [9].

5.1.2 Cardiovascular diseases

The inability to engage in regular aerobic activity due to physical disabilities increases susceptibility to cardiovascular diseases, such as coronary artery disease and stroke [10]. These risks are heightened by sedentary lifestyles and poor access to tailored exercise programs. Cardiovascular fitness is vital for enhancing stamina and reducing fatigue, especially in individuals relying on assistive devices for mobility [10].

5.1.3 Pressure sores

Physically disabled individuals who are confined mostly to bed are at risk of developing pressure sores secondary to circulatory impairment over specific skin areas with bony prominence. These sores can lead to severe infections and hospitalization, further reducing mobility level and independence. Effective prevention strategies include frequent repositioning, use of pressure-relieving cushions, and skin care management.

5.1.4 Musculoskeletal problems

Secondary musculoskeletal problems, such as joint deformities, osteoporosis, and contractures, are common in individuals with physical disabilities [11]. These issues arise from prolonged immobilization, muscle imbalance, or reliance on assistive devices, underscoring the need for early therapeutic interventions and strength training. Addressing secondary health conditions and mobility challenges is integral to improving the quality of life for individuals with physical disabilities. Multidisciplinary approaches that include tailored physical activity, rehabilitative care, and assistive technology are essential for mitigating these impacts. Through proactive management, individuals with physical disabilities can regain independence, reduce health risks, and lead fulfilling lives.

6. Reduced independence and mobility

Physical disabilities significantly impact an individual's independence and mobility, limiting their ability to perform daily tasks and navigate various environments. The effects of reduced mobility extend beyond physical challenges, influencing social participation, mental health, and overall quality of life. Restricted movement often necessitates reliance on caregivers for basic activities such as bathing, dressing, grooming, or meal preparation [11, 12]. This dependency can erode self-esteem and contribute to feelings of helplessness.

Limited mobility poses challenges in accessing education and job opportunities, leading to financial dependency and reduced social engagement [13]. For example, inaccessible workplaces or lack of transportation options can hinder career growth, impacting the economic stability of individuals with disabilities. Restricted independence often leads to diminished participation in community or social activities [14]. Loneliness and reduced social interaction are common, further exacerbating the psychological toll of physical disabilities.

7. Mental health challenges

Living with a physical disability often brings profound mental health challenges. The interplay between physical limitations and psychological well-being can lead to significant emotional distress, affecting an individual's ability to cope with their condition and maintain a positive quality of life.

7.1 Anxiety, depression, and psychological distress

Physical disabilities often lead to anxiety and depression due to the substantial changes they bring to daily life. Individuals may experience stress related to reduced ability to perform tasks, dependence on others, and fear of societal judgment or rejection [35]. Many individuals with physical disabilities report heightened anxiety due to fear of injury, loss of independence, or stigmatization [15]. This anxiety can become chronic, reducing overall resilience and engagement in rehabilitation efforts.

Depression is a common mental health condition among individuals with physical disabilities [16]. It is often linked to feelings of hopelessness, isolation, and frustration with their physical limitations. Studies show that people with disabilities are more likely to experience depression than the general population [17]. Chronic pain, fatigue, and the persistent struggle to adapt to a new reality can exacerbate psychological distress. This distress affects not only the individual but also their family and support system [18].

7.2 Body image issues

Living with a physical disability often alters an individual's perception of their body, leading to issues with body image and self-esteem [19, 36, 37]. Amputations, assistive devices, or visible deformities can lead to discomfort and self-consciousness [21]. This can result in individuals avoiding social interactions or public appearances. Unrealistic societal standards of beauty and fitness further compound body image issues. Individuals with physical disabilities may feel marginalized or judged based on their appearance. For those who acquire disabilities later in life, adapting to a new self-image can be particularly challenging [36, 37]. The transition often involves grieving the loss of their previous physical abilities.

7.3 Coping with disability

Developing healthy coping mechanisms is essential for addressing mental health challenges associated with physical disabilities. Strong social and emotional support from family, friends, and counselors is critical in helping individuals manage stress and build resilience [22, 38]. Counseling and psychotherapy, including cognitive-behavioral therapy (CBT), can help individuals navigate the emotional complexities of their disability. Engaging in adapted physical activities or recreational programs has been shown to improve mental health, reduce feelings of isolation, and foster a sense of accomplishment.

8. Social participation and isolation

Social participation is a vital aspect of well-being for individuals with physical disabilities. However, barriers to employment, education, and community engagement, coupled with mobility restrictions, can often lead to feelings of isolation and exclusion. Individuals with physical disabilities face systemic and environmental obstacles that hinder their ability to participate fully in societal activities [23]. Many individuals with disabilities encounter workplace discrimination, inaccessible work environments, or lack of accommodations, limiting their job opportunities [22].

Research indicates that employment rates for people with disabilities are significantly lower than those for people without disabilities, despite having comparable skills and qualifications [24]. Access to quality education can be restricted due to physical barriers in schools, lack of assistive technologies, or insufficient teacher training in inclusive education practices. This disparity often limits the professional and personal growth of individuals with physical disabilities. Social events, recreational activities, and cultural programs are often designed without considering accessibility. The lack of inclusive practices can marginalize individuals with disabilities, leaving them excluded from community life.

9. Neurobiological mechanisms

Physical activity influences brain function through key neurobiological mechanisms, including endorphin release, stress hormone regulation, and increased production of brain-derived neurotrophic factor (BDNF). These processes enhance mood, reduce anxiety, and promote neuroplasticity. Additionally, improved cerebral blood flow supports cognitive function, making exercise a vital tool for mental health and neurorehabilitation.

9.1 Endorphin release and mood regulation

Physical activity stimulates the release of endorphins, which are natural opioid peptides produced by the brain. These endorphins bind to opioid receptors, reducing pain perception and inducing feelings of pleasure and euphoria, often referred to as the "runner's high" [39]. This mechanism is particularly beneficial for individuals with physical disabilities, as it can alleviate chronic pain and improve overall mood. Studies have shown that even moderate-intensity exercise, such as wheelchair-assisted cycling or adaptive yoga, can trigger endorphin release, leading to significant reductions in symptoms of depression and anxiety [40–42].

9.2 Regulation of stress hormones

Exercise also plays a critical role in regulating the body's stress response. Physical activity reduces levels of cortisol, a hormone released during periods of stress, while simultaneously increasing the production of norepinephrine and dopamine, which help the brain manage stress more effectively [43]. For individuals with physical disabilities, who may face heightened stress due to social stigma or physical limitations, regular exercise provides a vital outlet for stress relief and emotional regulation.

9.3 Neuroplasticity and cognitive function

Another key benefit of physical activity is its impact on neuroplasticity, the brain's ability to adapt and form new neural connections. Exercise promotes the release of brain-derived neurotrophic factor (BDNF), a protein that supports the growth and survival of neurons, enhances synaptic plasticity, and improves cognitive functions such as memory, attention, and learning [44]. Adaptive physical activities, including seated Tai Chi or resistance band exercises, can stimulate BDNF production, contributing to better cognitive performance and emotional resilience, even in individuals with significant physical impairments.

10. Psychological mechanisms

Psychological support, such as counseling, peer support groups, and motivational coaching, is essential in helping individuals overcome these barriers and embrace physical activity as part of their rehabilitation and self-care routine. Awareness campaigns are also crucial in shifting societal attitudes toward disability and promoting inclusive participation in sports and physical activities.

10.1 Improved self-efficacy and confidence

Physical activity fosters a sense of achievement and mastery, which is crucial for improving self-efficacy—the belief in one's ability to succeed in specific tasks. For people with physical disabilities, achieving fitness goals, such as completing a set of adaptive exercises, can enhance self-confidence and promote a positive self-image [45]. This boost in self-efficacy can help individuals overcome feelings of helplessness and foster a proactive approach to managing their health and well-being.

10.2 Social connectedness and emotional support

Participating in group-based adaptive sports or exercise classes can also foster social interaction and a sense of belonging. These activities provide opportunities to connect with peers who share similar challenges, creating a supportive environment that reduces feelings of isolation and promotes emotional well-being [2]. Social support from family, friends, and caregivers further reinforces motivation and encourages sustained participation in physical activity.

10.3 Reduction in symptoms of depression and anxiety

Engaging in physical activity has been shown to reduce symptoms of depression and anxiety through both biological and psychological pathways. Exercise increases the availability of serotonin and dopamine, neurotransmitters associated with mood regulation [43]. Additionally, the distraction hypothesis suggests that focusing on physical movement during exercise provides a mental break from negative thought patterns, thereby reducing anxiety and promoting a more positive mental state.

11. Impact of mobility restrictions on relationships and community involvement

Mobility restrictions in individuals with physical disabilities also bring difficulties for them to socialize with friends and family, maintain relationships, and engage with their communities, thus resulting in feelings of loneliness [46]. Romantic relationships can also be impacted by the stigma surrounding disabilities and societal biases. Poor access or inaccessible public transportation, as well as unfriendly infrastructure and facilities, create obstacles for individuals with physical disabilities, hindering their participation in community activities. This isolation can perpetuate negative stereotypes and further alienate disabled individuals from society. Limited access to cultural venues, sports facilities, and public spaces can prevent individuals with disabilities from enjoying leisure activities, further narrowing their social world.

12. Mitigating isolation through inclusion and advocacy

Addressing social isolation requires coordinated efforts at both individual and societal levels. Governments and organizations must implement policies promoting accessibility in employment, education, and community programs. Public campaigns to reduce stigma and foster empathy can encourage greater acceptance of individuals with physical disabilities. Assistive devices and adaptive technologies can improve mobility and enable greater participation in various aspects of life.

13. Role of movement in empowerment for people with physical disabilities

Movement is a cornerstone for enhancing the lives of individuals with physical disabilities, serving as a catalyst for physical, mental, emotional, and social empowerment. Engaging in physical activity not only improves health outcomes but also fosters a sense of urgency and belonging, addressing the multifaceted challenges faced by this population.

13.1 Physical benefits

Regular physical activity offers profound physical advantages, enhancing overall health and preventing secondary complications [25]. Regular physical activity offers significant advantages to the physical health of individuals with disabilities. First, it enhances strength, endurance, and range of motion, crucial for mobility and daily activities. By maintaining and building muscle strength, individuals can perform tasks more efficiently and reduce dependency on others. Endurance-based exercises, such as hand cycling or swimming, improve cardiovascular health, allowing for sustained energy throughout the day. Additionally, range-of-motion exercises help preserve joint flexibility, preventing stiffness and reducing the risk of contractures.

A critical aspect of physical activity is its ability to prevent secondary complications. Prolonged inactivity often leads to pressure ulcers, particularly for individuals using wheelchairs. Regular movement promotes better blood circulation, reducing this risk. Similarly, inactivity-related muscle atrophy can be mitigated through strength training and consistent physical engagement. Furthermore, improved metabolism and circulation contribute to weight management, significantly lowering the risk of chronic conditions such as diabetes and heart disease. By addressing these physical challenges, regular movement enhances overall well-being and ensures better health outcomes. **Figure 3** shows the interplay between physical disabilities and mental health challenges on mental health.

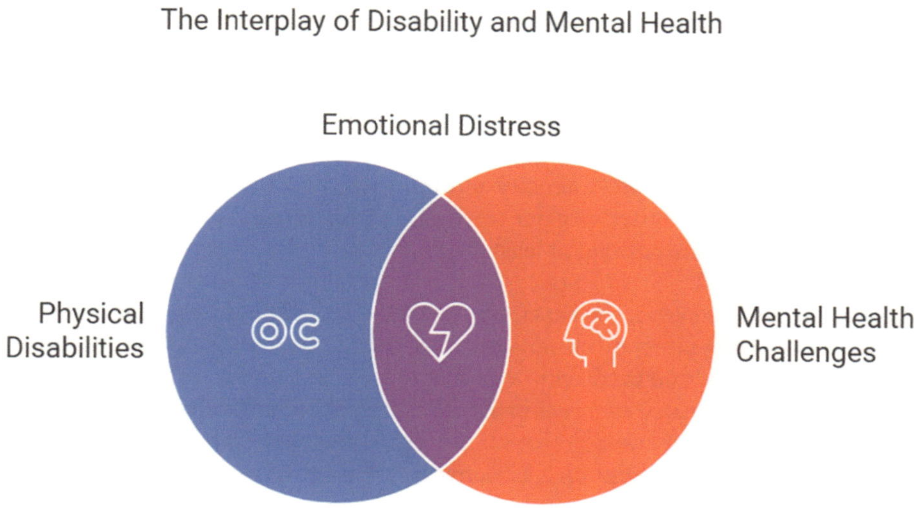

Figure 3.
The diagram shows the relationship between physical disabilities and mental health which results in emotional distress among PwD.

13.2 Mental and emotional benefits

The psychological impact of physical activity extends far beyond its physical benefits, playing a pivotal role in mental health and emotional resilience. Regular physical activity is a proven method for enhancing mood, reducing anxiety, and combating depression [25]. Engaging in movement stimulates the release of endorphins and serotonin, hormones that naturally improve mood and alleviate stress. Furthermore, physical activity introduces structure into daily routines, providing purpose and reducing feelings of aimlessness or disconnection.

Another significant emotional benefit of movement is its ability to increase self-esteem and confidence. Achieving fitness goals or mastering specific exercises fosters a profound sense of accomplishment. For individuals with physical disabilities, these milestones reaffirm their capabilities, empowering them to embrace their bodies and their potential. Physical activity also promotes a sense of autonomy and mastery over one's limitations, which can translate into greater confidence in facing other aspects of life [26].

13.3 Social benefits

Physical activity often creates opportunities for social inclusion, combating the isolation that individuals with physical disabilities may experience [27]. Participating in group exercises or adaptive sports fosters social interaction and teamwork, allowing individuals to connect with peers who share similar experiences. These connections are vital for mutual support, camaraderie, and the development of meaningful relationships. Moreover, community-based sports and rehabilitation programs play a crucial role in integrating individuals with disabilities into society [26]. Initiatives like wheelchair basketball, adaptive yoga, or inclusive fitness centers not only promote physical health but also serve as platforms for personal growth and skill development. These programs encourage participants to engage with their communities, fostering a sense of belonging and enabling them to contribute to collective activities.

14. Physical activity general guidelines

Physical activity is essential for people with physical disabilities to improve overall health, maintain mobility, and enhance quality of life. The World Health Organization (WHO) recommends that individuals with disabilities engage in at least 150 minutes of moderate-intensity aerobic physical activity or 75 minutes of vigorous-intensity aerobic activity per week [47]. Additionally, muscle-strengthening activities should be incorporated at least twice a week, tailored to the individual's ability and safety [28].

Adaptations to exercises are necessary to accommodate different physical disabilities. This includes ensuring safety by minimizing the risk of injury, ensuring accessibility to facilities or equipment, and modifying activities to suit the individual's functional capacity [29]. For example, seated exercises or wheelchair-specific workouts can be developed to suit people with limited lower-limb mobility. Gradual progression is key to avoiding overexertion and building endurance. Effective physical activity programs should focus on inclusivity, allowing individuals to engage in exercises that suit their preferences, abilities, and personal goals. Collaborative planning with healthcare professionals ensures these activities are both beneficial and enjoyable.

Consistency in physical activity is vital for individuals with physical disabilities to reap long-term benefits. Encouraging regular exercise routines helps maintain strength, endurance, and flexibility while preventing secondary complications like pressure sores or muscle atrophy. Setting achievable goals ensures sustained motivation and builds confidence in an individual's capabilities. Routines adapted to the individual's abilities and preferences promote adherence, making physical activity a positive, empowering experience.

15. Practical recommendations for healthcare professionals, enhancing physical activity for people with physical disabilities

Healthcare professionals play a critical role in promoting physical activity among individuals with physical disabilities. Tailored exercise programs and therapeutic protocols can improve physical function, reduce secondary health risks, and enhance mental well-being [48]. Below, we present comprehensive guidelines for healthcare professionals, focusing on specific exercises and therapeutic approaches for various disabilities.

15.1 Spinal cord injury (SCI)

Objectives: Improve cardiovascular health, enhance upper body strength, and reduce spasticity.

Recommended exercises:

- Wheelchair propulsion training: Improves cardiovascular endurance and upper body strength. Encourage interval training by alternating between slow and fast propulsion for 10–15 minutes, 3–4 times per week.

- Resistance band exercises: Strengthen upper body muscles. Focus on shoulder presses, bicep curls, triceps push, and seated rows, performing 2–3 sets of 10–12 repetitions, 2–3 times per week.

- Stretching protocols: Reduce spasticity and improve flexibility. Include seated hamstring and shoulder stretches, holding each for 20–30 seconds, 2–3 times daily.

Therapeutic protocols:

- Incorporate functional electrical stimulation (FES) cycling to stimulate paralyzed muscles and enhance cardiovascular fitness.

- Use body-weight-supported treadmill training (BWSTT) for incomplete SCI patients to improve gait and balance.

15.2 Cerebral palsy (CP)

Objectives: Improve motor control, reduce spasticity, and enhance balance.
Recommended exercises:

- Aquatic therapy: The buoyancy of water reduces joint stress and encourages free movement. Focus on water walking, leg lifts, and gentle swimming for 30–45 minutes, 2–3 times per week.

- Balance training: Use wobble boards or stability balls to improve postural control. Start with 5–10 minutes of static balance exercises, progressing to dynamic tasks such as reaching or throwing a ball.

- Strength training: Incorporate light weights or resistance bands for leg presses, seated rows, and lateral raises, performing 2–3 sets of 8–12 repetitions, 2 times per week.

Therapeutic protocols:

- Implement task-specific training, such as practicing stair climbing or transfers, to enhance functional mobility.

- Use constraint-induced movement therapy (CIMT) to improve motor function in the less-affected limb.

15.3 Stroke survivors

Objectives: Restore motor function, improve balance, and prevent secondary complications.
Recommended exercises:

- Gait training: Use parallel bars or a treadmill to improve walking ability. Start with 10–15 minutes of walking, gradually increasing speed and duration over time.

- Functional strength training: Include sit-to-stand exercises, heel raises, and step-ups to improve lower-limb strength and mobility. Perform 2–3 sets of 8–12 repetitions, 3 times per week.

- Coordination exercises: Incorporate tasks like finger-to-nose tapping or using a therapy ball to enhance hand-eye coordination.

Therapeutic protocols:

- Use mirror therapy to improve motor function by creating a visual illusion of normal movement in the affected limb.

- Apply robotic-assisted therapy for upper limb rehabilitation to facilitate repetitive, controlled movements.

15.4 Amputees

Objectives: Improve strength and balance, reduce the risk of falls, and facilitate prosthetic use.

Recommended exercises:

- Core strengthening: Include exercises like planks and seated leg lifts to enhance stability and support prosthetic use. Perform 2–3 sets of 10–15 repetitions, 2–3 times per week.

- Prosthetic gait training: Practice walking on various surfaces (e.g., grass, ramps, stairs) to improve adaptability and confidence. Start with short sessions, gradually increasing distance and difficulty.

- Balance training: Use single-leg stands or balance boards to improve postural control and reduce fall risk.

Therapeutic protocols:

- Utilize desensitization techniques, such as gentle massage and graded exposure to textures, to reduce phantom limb pain.

- Incorporate virtual reality (VR) training to simulate real-world environments and improve mobility.

16. Adaptive equipment and technology

Integrating adaptive equipment and technology has revolutionized physical activity for individuals with physical disabilities [25]. Devices like wheelchairs, prosthetics, and exoskeletons enable mobility and independence, making it easier to engage in regular movement. Assistive devices tailored to individual needs, such as custom orthotics or hand grips, allow for safe and effective participation in physical activity.

Technology, including virtual reality, adaptive gaming, and mobile apps, is becoming significant in promoting movement [23]. Virtual reality provides immersive experiences for low-impact exercise, while adaptive gaming fosters engagement through entertaining and movement-centered challenges. Mobile apps track progress and motivate users to remain consistent in their physical activity [30].

Programs like wheelchair basketball, adaptive swimming, and para-athletics allow individuals with physical disabilities to participate in sports, build confidence, and connect socially [30, 49]. These programs demonstrate how sport fosters empowerment and psychosocial well-being. Rehabilitation programs based in communities offer tailored support, ensuring that individuals with disabilities can engage in physical activity in familiar settings [50, 51]. Community initiatives often integrate cultural and local resources, promoting accessibility and sustainability.

17. Barriers to physical activity in people with physical disabilities

A lack of accessible infrastructure and unsafe public spaces often prevent individuals with disabilities from engaging in physical activity [31]. Inadequate sidewalks, lack of ramps, and inaccessible public transportation all contribute to physical barriers that hinder mobility. Furthermore, the design of public spaces often does not account for the diverse needs of people with disabilities, creating an environment where participation in community events, exercise, and recreation is limited [32]. Inclusive urban planning and infrastructure improvements are critical in overcoming these challenges. Cities and communities must prioritize universal design principles that ensure access for all, regardless of physical ability, and invest in creating spaces that facilitate mobility and social interaction for individuals with disabilities.

Adaptive equipment, rehabilitation services, and participation in sports programs often come with high costs, making them inaccessible to many individuals. Specialized wheelchairs, prosthetics, and assistive devices, as well as personalized rehabilitation programs, can be prohibitively expensive for those without adequate insurance or financial resources. Furthermore, some individuals may face challenges in accessing quality equipment and services in rural or underserved areas. Advocacy for funding, subsidies, and insurance coverage for assistive devices and rehabilitation services is essential to address these financial barriers [31, 52]. Government programs, charitable organizations, and public-private partnerships can play a pivotal role in making these services more affordable and accessible to individuals with physical disabilities.

Feelings of discouragement, lack of motivation, and societal stigma are additional psychological barriers that can deter individuals from participating in physical activity [53, 54]. Negative perceptions of disability, coupled with low expectations from society and even healthcare providers, can undermine an individual's confidence in their ability to engage in movement [55]. These mental barriers are often exacerbated by the lack of representation of people with disabilities in sports and physical activity contexts, which can perpetuate stereotypes and limit opportunities for inclusion [34]. Psychological support, such as counseling, peer support groups, and motivational coaching, is essential in helping individuals overcome these barriers and embrace physical activity as part of their rehabilitation and self-care routine. Awareness campaigns are also crucial in shifting societal attitudes toward disability and promoting inclusive participation in sports and physical activities.

Limited access to rehabilitation specialists and structured physical activity programs further restricts opportunities for individuals to engage in movement [40]. Many people with disabilities may not have access to trained professionals who specialize in adaptive physical therapy and inclusive rehabilitation practices. In some regions, there is a shortage of healthcare providers who are knowledgeable about the specific needs and abilities of individuals with disabilities, leading to gaps in care and treatment.

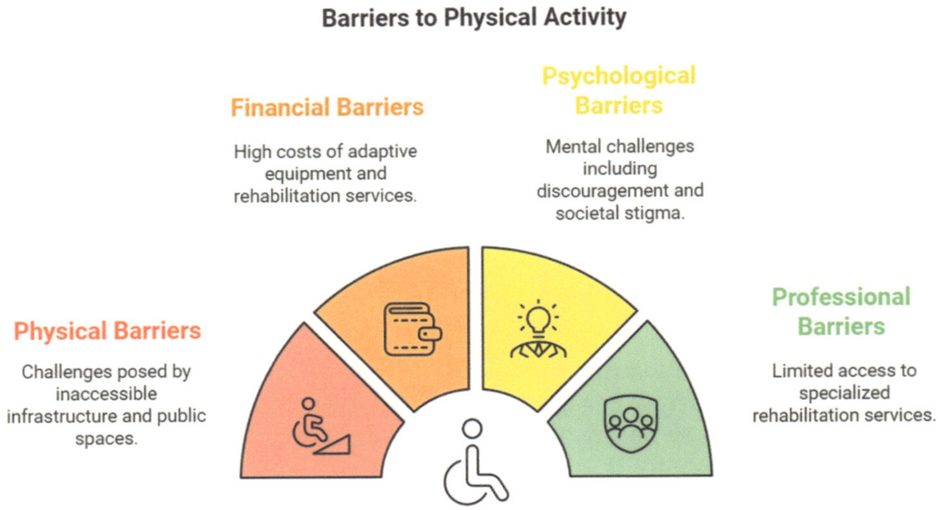

Figure 4.
The diagram shows the factors that can hinder PwD from doing physical activity.

Expanding healthcare services and training professionals in inclusive rehabilitation practice is essential to bridge this gap. This includes developing specialized programs in hospitals, rehabilitation centers, and community organizations that focus on adaptive physical activity and rehabilitation for individuals with physical disabilities. Training healthcare professionals to recognize the importance of movement for both physical and mental health, and providing them with the tools and knowledge to design inclusive and accessible exercise programs, can empower individuals with disabilities to take charge of their health and well-being.

By addressing these systemic barriers, whether physical, financial, psychological, or professional, society can create a more supportive and inclusive environment for individuals with disabilities. These efforts will enable people with disabilities to fully participate in physical activity, improving their health, quality of life, and social inclusion. **Figure 4** shows the barriers to physical activities among PwD.

18. Facilitators to physical activity in people with physical disabilities

Engaging in physical activity is a powerful tool for improving the quality of life and overall well-being of individuals with physical disabilities [56]. However, participation often hinges on the presence of facilitators that enable access, inclusion, and motivation [57]. These facilitators, encompassing accessible infrastructure, social support networks, adaptive technologies, and personal motivation, play a pivotal role in empowering individuals to overcome barriers and embrace movement [31, 32].

Accessible infrastructure is a fundamental facilitator of physical activity for people with physical disabilities [34]. Inclusive fitness centers equipped with adaptive exercise machines, wheelchair-accessible spaces, and trained staff create welcoming environments for individuals with diverse needs [40]. Accessible sports facilities, such as those designed for wheelchair basketball or rugby, provide safe and functional spaces for competitive and recreational activities [34]. Moreover, the development of inclusive community spaces, including parks with ramps, smooth pathways, and appropriate signage, ensures that people with disabilities can participate in outdoor activities alongside their peers. Such infrastructure fosters inclusivity and promotes a sense of belonging within communities.

Social support networks are equally critical in facilitating physical activity. Family members and caregivers often play a vital role by encouraging, assisting with transportation, and participating in activities to foster engagement. Peer groups offer a shared sense of camaraderie and motivation, as individuals with similar challenges inspire and support one another during group exercises or adaptive sports. Advocacy organizations further enhance participation by promoting awareness, organizing events, and advocating for resources to support physical activity for individuals with disabilities [58]. These networks not only boost participation but also contribute to emotional resilience and self-confidence.

The advent of adaptive equipment and advanced technologies has revolutionized the accessibility of physical activity for people with physical disabilities. Prosthetics and orthotics, for example, provide enhanced mobility and stability, enabling users to participate in various physical activities. Assistive devices, such as wheelchair-friendly treadmills, hand cycles, or exoskeletons, offer opportunities for tailored exercise regimens [33]. Additionally, technological innovations, such as virtual reality and adaptive gaming platforms provide engaging and low-impact ways for individuals to stay active. These tools allow users to explore new forms of movement and connect with others in virtual environments, making physical activity more accessible and enjoyable.

Last, personal motivation serves as a significant internal facilitator [59]. Setting achievable goals, celebrating milestones, and focusing on the mental and physical benefits of activity can inspire individuals to remain consistent in their efforts [60]. Rehabilitation specialists, such as physiotherapists and occupational therapists, play an instrumental role in fostering motivation by designing personalized movement programs that cater to individual abilities and preferences. Creating a positive and supportive atmosphere further encourages sustained participation.

In conclusion, facilitating physical activity for people with physical disabilities requires a multidimensional approach. By investing in accessible infrastructure, strengthening social support networks, leveraging adaptive technologies, and nurturing personal motivation, society can create an inclusive environment that empowers

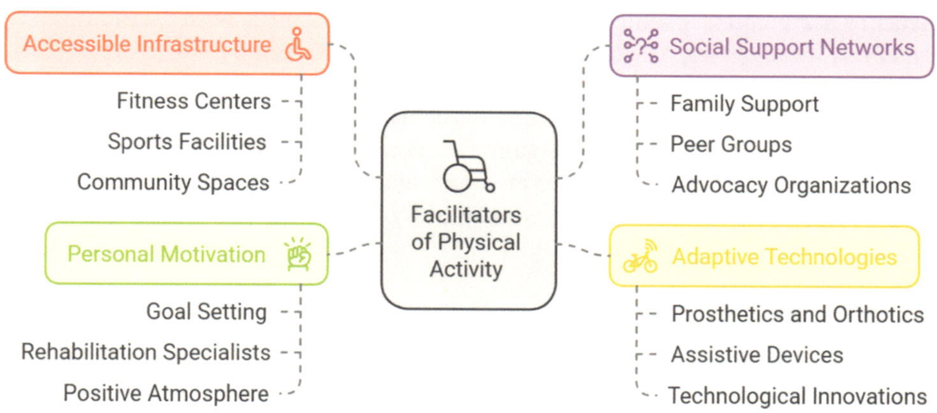

Figure 5.
The illustrations show the factors that can encourage PwD to do physical activity.

individuals with disabilities to lead active and fulfilling lives. These efforts not only enhance physical health but also promote social inclusion, mental well-being, and a greater sense of independence. **Figure 5** shows the facilitators of physical activity among PwD.

19. Overcoming barriers to physical activity for people with physical disabilities

Participation in physical activity offers numerous physical, mental, and social benefits for individuals with physical disabilities. However, many face significant barriers that hinder their ability to engage in such activities. These barriers can be categorized as physical, financial, psychological, and professional. Addressing these obstacles through inclusive strategies is essential to fostering an environment that encourages physical activity, enhances quality of life, and promotes social inclusion.

19.1 Physical barriers

A lack of accessible infrastructure and unsafe public spaces often restrict mobility and limit opportunities for physical activity among individuals with physical disabilities. Inadequate sidewalks, a shortage of ramps, and inaccessible public transportation contribute to these physical barriers [8]. Public spaces are often not designed with diverse needs in mind, creating environments that deter participation in community events, exercise, and recreation. To overcome these barriers, universal design principles must be implemented in urban planning, ensuring that spaces are accessible to everyone, regardless of physical ability. Cities should retrofit existing infrastructure with ramps, elevators, and tactile paving, as well as ensure that public transportation systems are fully accessible. For example, Barcelona, Spain, has made significant strides in accessibility, retrofitting older buildings and ensuring nearly all public transport is wheelchair-friendly [61].

19.2 Financial barriers

The cost of adaptive equipment, rehabilitation services, and participation in sports programs can be prohibitively high for many individuals with disabilities. Specialized wheelchairs, prosthetics, and assistive devices often require substantial financial resources, which may not be covered by insurance or accessible in rural areas. Advocacy for funding and subsidies is critical to making these services more affordable [62]. Governments and organizations can play a pivotal role by providing financial assistance through programs such as Australia's National Disability Insurance Scheme (NDIS), which funds disability-related support, including adaptive technologies and therapy [34].

19.3 Psychological barriers

Psychological barriers, such as lack of motivation, societal stigma, and low self-confidence, also deter individuals from participating in physical activity [57, 63]. Societal stigma and negative perceptions of disability can undermine confidence and perpetuate low expectations [41]. Psychological support, including counseling, peer support groups, and motivational coaching, can help individuals overcome these mental barriers and embrace physical activity as part of their rehabilitation. Awareness campaigns that highlight the achievements of individuals with disabilities in sports can also shift societal attitudes. The "We're The Superhumans" campaign for the 2016 Paralympics in Rio de Janeiro is a prime example, showcasing the capabilities of athletes with disabilities and challenging stereotypes [30].

19.4 Professional barriers

Limited access to trained rehabilitation specialists and structured physical activity programs further restricts opportunities for individuals with disabilities to engage in movement. In some regions, there is a shortage of healthcare providers knowledgeable about adaptive physical therapy and inclusive rehabilitation practices [28]. Expanding healthcare services and training professionals in inclusive rehabilitation is crucial. Programs that integrate tele-rehabilitation can also bridge the gap in rural and underserved areas. The Centre of Rehabilitation and Special Needs Studies (ICaRehab) initiative from the Faculty of Health Sciences, University Kebangsaan Malaysia in Malaysia, is an excellent example, offering community-based rehabilitation services while training physiotherapists in adaptive practices to ensure comprehensive care [64].

19.5 Social support and advocacy

Social support networks, including family, caregivers, and peer groups, play a critical role in encouraging physical activity. Advocacy organizations can further enhance participation by organizing inclusive sports events and promoting resources for individuals with disabilities. The Special Olympics, for instance, provides year-round sports training and competitions, fostering a sense of community and promoting inclusion [65].

Overcoming barriers to physical activity for people with physical disabilities requires a multifaceted approach. By addressing physical, financial [66], psychological, and professional obstacles, society can create a more inclusive environment that empowers individuals with disabilities to engage in movement and lead active, fulfilling lives. Such efforts not only improve physical health but also enhance mental well-being, social inclusion, and independence, contributing to a more equitable and inclusive society.

20. Emerging adaptive technologies and their practical applications

Emerging adaptive technologies, such as advanced prosthetics, exoskeletons, and virtual reality, are revolutionizing rehabilitation by enabling individuals with physical disabilities to regain mobility, enhance function, and improve independence.

20.1 Benefits of adaptive technologies in rehabilitation

Adaptive technologies significantly enhance rehabilitation outcomes by addressing physical, psychological, and social challenges faced by people with disabilities.

20.1.1 Improved physical function

Technologies like robotic exoskeletons and advanced prosthetics enable users to regain mobility, build strength, and improve coordination [67]. These benefits reduce the risk of secondary complications, such as joint stiffness and cardiovascular decline.

20.1.2 Increased therapy adherence

Gamified VR systems and engaging exercise equipment motivate users to stay consistent with their rehabilitation programs, leading to better long-term outcomes [68].

20.1.3 Enhanced independence and confidence

By enabling users to perform daily tasks and navigate their environments independently, adaptive technologies boost self-esteem and foster a sense of empowerment [69, 70].

20.1.4 Social inclusion and engagement

Adaptive sports equipment and inclusive fitness centers encourage participation in community activities, promoting social interaction and reducing isolation [71].

20.2 Advanced prosthetics and orthotics

Modern prosthetics, such as myoelectric limbs and dynamic-response foot prosthetics, mimic natural movements and improve functional mobility. Myoelectric prosthetics use muscle-generated electrical signals to perform tasks like gripping objects, which enhances independence in daily activities. Dynamic-response foot prosthetics store and return energy during movement, improving walking efficiency and enabling participation in sports. For example, Ottobock's *C-Leg 4*, a microprocessor-controlled knee, adapts to various terrains, giving users greater mobility and confidence in diverse environments.

20.3 Robotic exoskeletons for mobility assistance

Robotic exoskeletons provide mobility support for individuals with spinal cord injuries and neurological conditions. These devices enable users to stand and walk, improving muscle strength, joint mobility, and cardiovascular health. In rehabilitation settings, exoskeletons facilitate gait retraining by providing consistent, repetitive movement patterns that enhance neuroplasticity and motor learning. The psychological benefits of standing and walking, such as improved self-esteem and social engagement, are equally significant.

20.3.1 Virtual reality (VR) and augmented reality (AR)

VR and AR technologies create immersive and interactive environments that motivate individuals to engage in therapeutic exercises. VR systems simulate real-world scenarios, such as navigating crowded streets or balancing on uneven surfaces, which improve balance and spatial awareness. AR overlays digital instructions onto real-world settings, guiding users through specific movements and providing real-time feedback [68]. The *Neofect Smart Glove* exemplifies this technology by combining VR and motion sensors to support hand rehabilitation for stroke survivors through gamified exercises that enhance engagement and adherence [72].

20.3.2 Wearable sensors and smart devices

Wearable devices such as smartwatches, fitness trackers, and smart insoles provide real-time feedback on physical activity, heart rate, and movement patterns. These tools are particularly useful in home-based rehabilitation, where they allow healthcare professionals to monitor progress remotely [73]. Wearable technologies also encourage users to set and achieve fitness goals, promoting long-term adherence to physical activity routines.

20.3.3 Adaptive exercise equipment

Inclusive exercise equipment, such as wheelchair-accessible treadmills, hand cycles, and resistance machines, facilitates safe and effective physical activity for individuals with mobility impairments. These devices enable users to perform cardiovascular, strength, and flexibility exercises in both rehabilitation and community fitness settings [27]. For instance, the *SciFit PRO1 Upper Body Ergometer* allows wheelchair users to build upper body strength and endurance, making it a valuable tool for improving functional independence and overall fitness [74].

21. Challenges and future directions

Despite their transformative potential, adaptive technologies face several challenges, including high costs, limited availability in rural areas, and the need for specialized training for healthcare professionals. To address these challenges, governments and organizations must invest in the following:

21.1 Affordability and accessibility

Public-private partnerships and government subsidies can reduce the cost of adaptive technologies, ensuring their accessibility to individuals from all socioeconomic backgrounds.

21.2 Training and education

Comprehensive training programs for healthcare providers are essential to ensure effective implementation and optimal use of these technologies in rehabilitation.

21.3 Continued research and development

Ongoing innovation is necessary to create affordable, user-friendly technologies that address the diverse needs of people with disabilities. Adaptive technologies and innovations have transformed the rehabilitation landscape, enabling individuals with physical disabilities to lead more active, fulfilling lives. From advanced prosthetics and exoskeletons to VR-based therapies and wearable devices, these tools offer practical solutions for overcoming physical limitations and fostering inclusion in physical activity. By integrating these technologies into rehabilitation settings and addressing existing challenges, healthcare professionals can empower individuals with disabilities to achieve their full potential and contribute to a more inclusive society.

22. Future directions and innovations

Emerging technologies, such as robotics, artificial intelligence, and virtual reality, hold immense potential for improving mobility and rehabilitation outcomes [75]. Robotic devices and AI-driven programs can tailor exercises to individual needs, ensuring precision and effectiveness in rehabilitation [76]. These technologies can also adapt to varying levels of ability, providing a scalable approach to physical activity. Virtual reality (VR) introduces immersive environments that not only make exercises engaging but also help overcome psychological barriers by simulating real-world scenarios in a controlled and safe manner [77].

In addition to these advancements, wearable devices, such as smart sensors and fitness trackers, are becoming increasingly valuable in monitoring progress and offering real-time feedback [30]. These tools empower individuals to track their physical activity levels, set achievable goals, and celebrate milestones, fostering motivation and adherence to exercise routines. The integration of such technologies into rehabilitation programs can significantly enhance the personalization and overall impact of physical activity interventions.

Continued research into the long-term benefits of physical activity for individuals with physical disabilities is crucial. Future studies should focus on understanding the multifaceted effects of movement on mental health, social inclusion, and overall quality of life. Investigations into how physical activity influences neuroplasticity, functional recovery, and psychological resilience will provide deeper insights for refining intervention strategies. This research will also help identify the most effective modalities of exercise for specific disabilities, ensuring optimized outcomes for individuals.

Furthermore, global collaborations are essential to advancing accessibility in sports and rehabilitation. International efforts can standardize practices, share innovative technologies, and create inclusive platforms for participation. Initiatives aimed at developing universally accessible sports facilities, adaptive equipment, and inclusive policies will play a pivotal role in breaking down barriers to physical activity [33]. Programs that promote cultural sensitivity and address regional challenges will also ensure equitable access to structured physical activity opportunities worldwide.

By embracing technological advancements, fostering inclusive practices, and prioritizing rigorous research, society can pave the way for transformative changes in the lives of individuals with physical disabilities. These combined efforts will not only improve mobility and rehabilitation outcomes but also empower individuals to achieve greater independence, participate fully in their communities, and lead healthier, more fulfilling lives.

23. The transformative impact of physical activity: Case studies of individuals with physical disabilities

Physical activity has the potential to transform the lives of individuals with physical disabilities, not only by improving physical health but also by enhancing mental well-being and social inclusion. To illustrate these benefits, the following case studies highlight how movement-based interventions have positively impacted individuals across different contexts.

23.1 Case study 1: Bonnie Bunyau Gustin – Paralympic gold medalist in powerlifting

Bonnie Bunyau Gustin, a 24-year-old para-powerlifter from Sarawak, made history at the 2020 Tokyo Paralympic Games by winning the gold medal in the men's 72 kg category. Born with short stature, Bonnie initially struggled with societal perceptions of his disability. However, through his dedication to sports and the support of the National Sports Council of Malaysia, he discovered his passion for weightlifting. Bonnie's rigorous training and mental resilience exemplify how physical activity can empower individuals to overcome barriers and achieve greatness. His gold medal win not only brought pride to Malaysia but also inspired many individuals with disabilities to pursue sports and physical activity as a means of empowerment and inclusion. Bonnie's journey highlights the role of national programs and expert coaching in nurturing talent and breaking stereotypes (Paralympic Games, 2021).

23.2 Case study 2: Mohmad Ridzuan Mohmad Puzi – Paralympic gold medalist in athletics

Mohamad Ridzuan Mohamad Puzi, affectionately known as "Dek Wan," is a 35-year-old para-athlete with cerebral palsy who has overcome immense physical and psychological challenges to become a global icon. At the 2016 Rio Paralympic Games, Ridzuan clinched Malaysia's first-ever Paralympic gold medal in the men's 100 m T36 sprint, setting a Paralympic record. His remarkable achievement earned him the title of "Fastest Man in the World" in his category. Ridzuan's story is one of determination, resilience, and the transformative power of physical activity. Through consistent training and support from his coaches and family, he improved not only his physical capabilities but also his mental fortitude. His success has raised awareness about the potential of individuals with disabilities and has been a catalyst for greater investment in Malaysia's para-sports programs. Ridzuan's journey demonstrates the role of sports in fostering self-confidence, societal acceptance, and national pride (Bernama, 2016).

23.3 Case study 3: Abdul Latif Romly – Paralympic gold medalist in long jump

Abdul Latif Romly, a para-athlete with an intellectual disability, is another Malaysian Paralympic star who has made headlines globally. At the 2016 Rio Paralympic Games, Latif won the gold medal in the men's T20 long jump, breaking the world record three times in a single event. He repeated his success at the 2020 Tokyo Paralympic Games, securing another gold medal and cementing his status as one of the world's greatest para-athletes. Latif's journey began in Perlis, where he discovered his talent for the long jump through school sports programs. With support from his coaches and the National Sports Council of Malaysia, he refined his technique and mental preparedness. Physical activity not only enhanced Latif's athletic performance but also helped him gain confidence and develop life skills, such as discipline and perseverance. His achievements inspire many individuals with disabilities to pursue their potential, proving that limitations can be overcome with determination and the right support (Paralympic Games, 2021).

23.4 Case study 4: Cheah Liek Hou – Paralympic gold medalist in badminton

Cheah Liek Hou, a 34-year-old para-badminton player from Kuala Lumpur, made history at the 2020 Tokyo Paralympic Games by becoming the first-ever gold medalist in men's singles badminton (SU5 category), since the sport was included in the Paralympic Games. Born with Erb's palsy, which affects the movement of his right shoulder and arm, Cheah began playing badminton at a young age and quickly developed a passion for the sport. Cheah's journey to Paralympic success involved rigorous training, mental discipline, and a relentless drive to prove his abilities. His victory in Tokyo was a testament to his hard work and dedication, as well as the growing prominence of para-badminton on the global stage. Cheah's success has inspired many aspiring athletes with disabilities and raised awareness about the potential of adaptive sports in fostering inclusion and excellence.

Author details

Hafifi Hisham*, Mohd Fahmi Aluwi and Nor Azlin Mohd Nordin
Physiotherapy Program, Centre for Rehabilitation and Special Needs Studies, Faculty of Health, Sciences, Universiti Kebangsaan Malaysia, Kuala Lumpur, Malaysia

*Address all correspondence to: hafifi.hisham@ukm.edu.my

References

[1] Ginis KAM, van der Ploeg HP, Foster C, Lai B, McBride CB, Ng K, et al. Participation of people living with disabilities in physical activity: A global perspective. Lancet. 2021;**398**(10298):443-455

[2] Belcher BR, Zink J, Azad A, Campbell CE, Chakravartti SP, Herting MM. The roles of physical activity, exercise, and fitness in promoting resilience during adolescence: Effects on mental well-being and brain development. Biological Psychiatry: Cognitive Neuroscience and Neuroimaging. 2021;**6**(2):225-237

[3] Izquierdo M, Singh MF. Promoting resilience in the face of ageing and disease: The central role of exercise and physical activity. Ageing Research Reviews. 2023;**88**:101940

[4] WHO. Global Report on Health Equity for Persons with Disabilities. Geneva: World Health Organization; 2022

[5] Lo J, Chan L, Flynn S. A systematic review of the incidence, prevalence, costs, and activity and work limitations of amputation, osteoarthritis, rheumatoid arthritis, back pain, multiple sclerosis, spinal cord injury, stroke, and traumatic brain injury in the United States: A 2019 update. Archives of Physical Medicine and Rehabilitation. 2021;**102**(1):115-131

[6] Hashemi G, Wickenden M, Bright T, Kuper H. Barriers to accessing primary healthcare services for people with disabilities in low and middle-income countries, a meta-synthesis of qualitative studies. Disability and Rehabilitation. 2022;**44**(8):1207-1220

[7] James SL, Lucchesi LR, Bisignano C, Castle CD, Dingels ZV, Fox JT, et al. Morbidity and mortality from road injuries: Results from the global burden of disease study 2017. Injury Prevention. 2020;**26**(Suppl 2):i46-i56

[8] Bickenbach J. The world report on disability. Disability & Society. 2011;**26**(5):655-658

[9] Falola T, Hamel N. Disability in Africa: Inclusion, Care, and the Ethics of Humanity. Vol. 91. United Kingdom: Boydell & Brewer; 2021

[10] Mfoafo-M'Carthy M, Grischow J, Moasun F. Experiencing Disability Stigma in Ghana. London: Lexington Books; 2024

[11] Sharma A, Dunay A. "Disability isn't contagious, ignorance is": A critical overview on the various shade of disability. In: Economic and Social Development: Book of Proceeding; Warsaw, Poland. 2016

[12] Sherry K. Disability and rehabilitation: Essential considerations for equitable, accessible and poverty-reducing health care in South Africa. South African Health Review. 2014;**2014**(1):89-99

[13] O'Haver K. Youth with disabilities in Africa: Bridging the disability divide. In: The Routledge Companion to Disability and Media. New York: Routledge; 2019. pp. 189-198

[14] Kincaid K. Towards a Critical Global Disability Studies: A Case Study of the African Youth with Disabilities Network. UK: Thesis for MSc in International Development, University of Bath; 2013

[15] Oswin N. Global City Futures: Desire and Development in Singapore. Vol. 44.

Georgia: University of Georgia Press; 2019

[16] Watson V. 'The planned city sweeps the poor away...': Urban planning and 21st century urbanisation. Progress in Planning. 2009;**72**(3):151-193

[17] Karki J, Rushton S, Bhattarai S, De Witte L. Access to assistive technology for persons with disabilities: A critical review from Nepal, India and Bangladesh. Disability and Rehabilitation. Assistive Technology. 2023;**18**(1):8-16

[18] Dhakal R, Groves CC. Rehabilitation in Nepal. Physical Medicine and Rehabilitation Clinics of North America. 2019;**30**(4):787-794

[19] Farrar M, Niraula YR, Pryor W. Improving access to prosthetic services in Western Nepal: A local stakeholder perspective. Disability and Rehabilitation. 2023;**45**(7):1229-1238

[20] Manaf H, Shaid A, Justine M, Hisham H, Hasnan N, Mohd Asmawi UM, et al. Barriers to physical activity and exercise amongst persons with physical disability in a government-funded teaching hospital, Kuala Lumpur. Malaysian Journal of Medicine and Health Sciences. 2021;**18**:365-373

[21] Yukishita T. Promotion of sports for people with disabilities implemented by the Japan Sports Agency. Juntendo Medical Journal. 2018;**64**(6):430-436

[22] Ghosh D, Chatterjee K. India's present stand point and future possibilities regarding paralympics. International Journal of Physiology, Exercise and Physical Education. 2023;**5**(1):20-22

[23] Buhalis D, Darcy S, Ambrose I. Best Practice in Accessible Tourism: Inclusion, Disability, Ageing Population and Tourism. Vol. 53. UK, USA, Canada: Channel View Publications; 2012

[24] Bruntlett M, Bruntlett C. Building the Cycling City: The Dutch Blueprint for Urban Vitality. Washington: Island Press; 2018

[25] Hamraie A. Building Access: Universal Design and the Politics of Disability. Minneapolis: University of Minnesota Press; 2017

[26] Vivanti D. Persons with disabilities and the right to equality before the law: Contribution from the European Disability Forum. European Yearbook of Disability Law. 2014;**5**:151

[27] Hardin R, Pate JR. Introduction to Adaptive Sport and Recreation. U.S.A: Human Kinetics; 2024

[28] Rimmer JH, Marques AC. Physical activity for people with disabilities. Lancet. 2012;**380**(9838):193-195

[29] Mauerberg-deCastro E, Campbell DF, Tavares CP. The global reality of the paralympic movement: Challenges and opportunities in disability sports. Motriz. Revista de Educacao Fisica. 2016;**22**:111-123

[30] De Souza DL, Brittain I. The Rio 2016 paralympic games: The visibility of people with disabilities in Brazil as a possible legacy. Communication & Sport. 2022;**10**(2):334-353

[31] van Dijk A, Dawson E, Moore KM, Belanger P. Risk assessment during the Pan American and Parapan American games, Toronto, 2015. Public Health Reports. 2017;**132**(1_suppl):106S-110S

[32] Humpage L. Models of disability, work and welfare in Australia.

Social Policy and Administration. 2007;**41**(3):215-231

[33] Fortune N, Madden RH, Clifton S. Health and access to health services for people with disability in Australia: Data and data gaps. International Journal of Environmental Research and Public Health. 2021;**18**(21):11705

[34] Purcal C, Fisher KR, Meltzer A. Social insurance for individualized disability support: Implementing the Australian National Disability Insurance Scheme (NDIS). In: Social Policy Review. Vol. 28. U.K: Policy Press; 2016. pp. 173-190

[35] Tiwari S, Joshi A, Rai N, Satpathy P. Impact of stroke on quality of life of stroke survivors and their caregivers: A qualitative study from India. Journal of Neurosciences in Rural Practice. 2021;**12**(4):680

[36] Sarroca N, Valero J, Deus J, Casanova J, Luesma MJ, Lahoz M. Quality of life, body image and self-esteem in patients with unilateral transtibial amputations. Scientific Reports. 2021;**11**(1):12559

[37] Ward Khan Y, O'Keeffe F, Nolan M, Stow J, Davenport J. "Not a whole woman": An interpretative phenomenological analysis of the lived experience of women's body image and sexuality following amputation. Disability and Rehabilitation. 2021;**43**(2):251-261

[38] Lestari R, Fajar M. Social support and self-esteem in people with physical disabilities. Indigenous Jurnal Ilmiah Psikologi. 2020;**5**(2):207-217

[39] Winiarz E. Endorphins, endocannabinoids and runners' high. The Science Journal of the Lander College of Arts and Sciences. 2019;**13**(1):4

[40] Lowe DA, Wu N, Rohdin-Bibby L, Moore AH, Kelly N, Liu YE, et al. Effects of time-restricted eating on weight loss and other metabolic parameters in women and men with overweight and obesity: The TREAT randomized clinical trial. JAMA Internal Medicine. 2020;**180**(11):1491-1499

[41] Scully D, Kremer J, Meade MM, Graham R, Dudgeon K. Physical exercise and psychological well being: A critical review. British Journal of Sports Medicine. 1998;**32**(2):111-120

[42] Kim DI, Lee J, Park H, Jeon JY. The relationship between physical activity levels and mental health in individuals with spinal cord injury in South Korea. International Journal of Environmental Research and Public Health. 2020;**17**(12):4423

[43] Azizi SA. Monoamines: Dopamine, norepinephrine, and serotonin, beyond modulation,"switches" that alter the state of target networks. The Neuroscientist. 2022;**28**(2):121-143

[44] Rosley N, Hisham H, Amin NIM, Abd HS, Ghani HM, Nordin NAM. Brain-derived neurotrophic factor (BDNF) responses to exercise in individuals with spinal cord injury: A systematic review. Asian Journal of Medicine & Health Sciences. 2024;**7**(2):4

[45] Drandrov GL, Syukiev DN, Burlykov VD, Burlykova EB, Marsunov SN. Forming a positive self-image in adolescents with disabilities when playing sports. European Proceedings of Social and Behavioural Sciences. 2021;**107**:387-398

[46] Battalio SL, Huffman SE, Jensen MP. Longitudinal associations between physical activity, anxiety, and depression in adults with long-term

physical disabilities. Health Psychology. 2020;**39**(6):529

[47] Bull FC, Al-Ansari SS, Biddle S, Borodulin K, Buman MP, Cardon G, et al. World Health Organization 2020 guidelines on physical activity and sedentary behaviour. British Journal of Sports Medicine. 2020;**54**(24):1451-1462

[48] Hisham H, Justine M, Hasnan N, Manaf H. Effects of paraplegia fitness integrated training on physical function and exercise self-efficacy and adherence among individuals with spinal cord injury. Annals of Rehabilitation Medicine. 2022;**46**(1):33-44

[49] Tabacof L, Dewil S, Herrera JE, Cortes M, Putrino D. Adaptive esports for people with spinal cord injury: New Frontiers for inclusion in mainstream sports performance. Frontiers in Psychology. 2021;**12**:612350

[50] Isidoro-Cabañas E, Soto-Rodríguez FJ, Morales-Rodríguez FM, Pérez-Mármol JM. Benefits of adaptive sport on physical and mental quality of life in people with physical disabilities: A meta-analysis. Healthcare (Switzerland). 2023;**11**(18):2480

[51] Declerck L, Stoquart G, Lejeune T, Vanderthommen M, Kaux JF. Barriers to development and expansion of adaptive physical activity and sports for individuals with a physical disability in sports clubs and centres. Science & Sports. 2021;**36**(3):202-209

[52] Hisham H, Rosley H, Manaf H, Justine M. Barriers and facilitators to physical activity and exercise among individuals with spinal cord injury: A systematic review. Malaysian Journal of Medicine and Health Sciences. 2022

[53] Olasagasti-Ibargoien J, Castañeda-Babarro A, León-Guereño P, Uria-Olaizola N. Barriers to physical activity for women with physical disabilities: A systematic review. Journal of Functional Morphology and Kinesiology. 2023;**8**(2):82

[54] Ascondo J, Martín-López A, Iturricastillo A, Granados C, Garate I, Romaratezabala E, et al. Analysis of the barriers and motives for practicing physical activity and sport for people with a disability: Differences according to gender and type of disability. International Journal of Environmental Research and Public Health. 2023;**20**(2):1320

[55] Moran TE, Taliaferro AR, Pate JR. Confronting physical activity programming barriers for people with disabilities: The empowerment model. Quest. 2014;**66**(4):396-408

[56] Diaz R, Miller EK, Kraus E, Fredericson M. Impact of adaptive sports participation on quality of life. Sports Medicine and Arthroscopy Review. 2019;**27**(2):73-82

[57] Ives B, Clayton B, Brittain I, Mackintosh C. 'I'll always find a perfectly justified reason for not doing it': Challenges for disability sport and physical activity in the United Kingdom. Sport in Society. 2021;**24**(4):588-606

[58] Mohajan D, Mohajan HK. Obesity and its related diseases: A new escalating alarming in global health. Journal of Innovations in Medical Research. 2023;**2**(3):12-23

[59] Franklin BA, Thompson PD, Al-Zaiti SS, Albert CM, Hivert MF, Levine BD, et al. Exercise-related acute cardiovascular events and potential deleterious adaptations following long-term exercise training: Placing the risks into perspective–An update: A scientific statement from the American

Heart Association. Circulation. 2020;**141**(13):e705-e736

[60] Pandey S, Chouksey A, Pitakpatapee Y, Srivanitchapoom P. Movement disorders and musculoskeletal system: A reciprocal relationship. Movement Disorders Clinical Practice. 2022;**9**(2):156-169

[61] Monforte J, Úbeda-Colomer J, Pans M, Pérez-Samaniego V, Devís-Devís J. Environmental barriers and facilitators to physical activity among university students with physical disability-A qualitative study in Spain. International Journal of Environmental Research and Public Health. 2021, 2021;**18**(2):464

[62] Lysaght R, Fabrigar L, Larmour-Trode S, Stewart J, Friesen M. Measuring workplace social support for workers with disability. Journal of Occupational Rehabilitation. 2012;**22**:376-386

[63] Budd MA, Gater DR Jr, Channell I. Psychosocial consequences of spinal cord injury: A narrative review. Journal of Personalized Medicine. 2022;**12**(7):1178

[64] Ong W, Omar NA, Zanudin A, Alias MF, Wen LH, Ee ATX, et al. The relationship between physical activity and mental health among individuals with spinal cord injury: Protocol for a scoping review. JMIR Research Protocols. 2024;**13**(1):e56081

[65] Begidov MV, Begidova TP, Bespalova VV. The importance of adaptive sports for the social adaptation of persons with disabilities. Tambov University Review. Series Humanities. 2023;**1**(1):87-90

[66] Manaf H, Tahir I, Hisham H, Justine M, Hasnan N. Return to work after spinal cord injury: A systematic review. Advanced Science Letters. 2018;**24**(4):2374-2377

[67] Ali WARW, Kassim AA, Rashid SMM. Assistive technology for persons with physical disabilities: A literature review. International Journal of Academic Research in Business and Social Sciences. 2024;**14**(2):544-560

[68] Asadzadeh A, Samad-Soltani T, Salahzadeh Z, Rezaei-Hachesu P. Effectiveness of virtual reality-based exercise therapy in rehabilitation: A scoping review. Informatics in Medicine Unlocked. 2021;**24**:100562

[69] Rayes R, Ball C, Lee K, White C. Adaptive sports in spinal cord injury: A systematic review. Current Physical Medicine and Rehabilitation Reports. 2022;**10**(3):145-153

[70] Pousada Garcia T, Garabal-Barbeira J, Porto Trillo P, Vilar Figueira O, Novo Díaz C, Pereira Loureiro J. A framework for a new approach to empower users through low-cost and do-it-yourself assistive technology. International Journal of Environmental Research and Public Health. 2021;**18**(6):3039

[71] Martin Ginis KA, van der Ploeg HP, Foster C, Lai B, McBride CB, Ng K, et al. Participation of people living with disabilities in physical activity: A global perspective. The Lancet. 2021;**398**(10298):443-455

[72] Lansberg MG, Legault C, MacLellan A, Parikh A, Muccini J, Mlynash M, et al. Home-based virtual reality therapy for hand recovery after stroke. PM & R: The Journal of Injury, Function, and Rehabilitation. 2022;**14**(3):320-328

[73] Allen T, Shepherd J, Wood J, Tyler D, Duncan O. Wearables for disabled and extreme sports. In: Digital Health:

Exploring Use and Integration of Wearables. Elsevier Masson s.r.l.; 2021

[74] Holm NJ, Møller T, Schou LH, Biering-Sørensen F. Monitoring outcome measures for cardiometabolic disease during rehabilitation and follow-up in people with spinal cord injury. Spinal Cord. 2024;**62**(3):125-132

[75] Sumner J, Lim HW, Chong LS, Bundele A, Mukhopadhyay A, Kayambu G. Artificial intelligence in physical rehabilitation: A systematic review. Artificial Intelligence in Medicine. 2023;**146**:102693

[76] Jeyathevan G, Catharine Craven B, Cameron JI, Jaglal SB. Facilitators and barriers to supporting individuals with spinal cord injury in the community: Experiences of family caregivers and care recipients. Disability and Rehabilitation. 2020;**42**(13):1844-1854

[77] Ferdiana A, Post MWM, Bültmann U, Van Der Klink JJL. Barriers and facilitators for work and social participation among individuals with spinal cord injury in Indonesia. Spinal Cord. 2021;**59**(10):1079-1087

Chapter 8

Physical Activities Averting Women Demise

Hifza Ahmed, Haleema Sadia, Ulvina Riaz and Irtasam Ahmad

Abstract

The word physical activity primarily refers to any voluntary movement of the body that leads to energy expenditure. These movements can procure quite physical benefits along with social and intellectual if performed in a strategic manner. Fascinatingly, it is effects are more pronounced in females influenced by biological, psychological and socio-cultural factors. Rather than monotonous physical activities like proper exercise, women can involve themselves in more fun activities, namely, bicycling, hiking, swimming or simply playing with kids. Additionally, exercises such as aerobic or strength training for physical strengthening or yoga, meditation and tai chi for tranquility are also becoming popular among women. These activities are not only fun but also create a sense of fullness, self-love, self-satisfaction and joy in them, further combating issues like depression, anxiety and social isolation. Conclusively, to garner all these benefits among women, they need to enjoy it wholly, whether it is exercise or recreational activity.

Keywords: energy metabolism, social behavior, resistance training, mental health, mobility

1. Introduction

> *"Physical Activity is any movement of the body that requires energy expenditure. This includes voluntary movements, such as exercise and daily activities, and involuntary movements, like those that occur during digestion and posture maintenance"* [1].

Fundamentally, any movement of the body that uses energy is considered physical exercise. This broad category encompasses both voluntary movements (acts that are done according to one's own volition) and involuntary movements (automatic body functions). Voluntary movements include exercising (organized physical activities aimed at increasing fitness) or going about daily activities (physically demanding daily tasks), such as strolling or cultivating garden. While involuntary bodily movements include many body functions ranging from simple to complex ones like the contractions needed to facilitate digestion (muscle contractions) or the constant posture corrections (modifying muscles and bones to maintain balance and uprightness). In short, physical activity, which includes a variety of activities that support our general health and well-being, is essentially a basic component of human existence [2].

Now, let us delve a little into the fuel or the energy required to perform physical activity. Interestingly, the ability to measure energy expenditure during physical activity in individuals is crucial. It is also critical to recognize that physical activity and energy expenditure are two distinct concepts, affecting one another in certain ways [3]. An idea is that the basic cellular repair and maintenance, digestion, the regulation of temperature, movement, growth, and reproduction are just a few of the ways through which people consume energy. Their daily energy expenditure may fluctuate significantly as a result of these processes' leading to notable shift. For instance, it has been assumed that activity and daily energy expenditure are lower during simple daily activities for example during locomotory maneuver since the amount of locomotion required for these activities is frequently relatively modest. Furthermore, spontaneous physical activity and non-exercise activities have received a lot of attention. The latter is frequently defined so broadly as comprising any energy expended as a result of activity, excluding voluntary exercise. In this case, physical activity and energy expenditure do have a dependence on each other [4].

In the following chapter, the possible effects of different types of physical activities whether exercise or recreational activities particularly on the women will be discussed. Along with that, some new theories and innovations regarding the use of physical activities among women will also be reviewed.

1.1 Physical activity: A must have concept in life

Physical activity is important for everyone and simply talking about the effect of normal physical activity on any person, its benefits are just immense. To maintain a healthy body by improving the normal muscular skeleton system along with cardiovascular health, performing a regular amount of physical activity is deemed necessary for all individuals [5]. Interestingly, discussing its impact not only on just physical health; these kinds of physical activities can also affect positively a normal individual's cognition and memory level [6]. All of this simply leads to a decrease in anxiety and depression levels while improving self-confidence in people. Fascinatingly, an active lifestyle of someone starting specifically from childhood to late adulthood can considerably improve one's quality of life and further decrease the risk of many kinds of chronic diseases as the person ages [7].

Even the World Health Organization (WHO) has recommended that children starting at the age of five and moving to the youth of 17 years old should at least involve themselves in 60 minutes of physical activity from moderate to intense. It could be anything like a brisk walk, playing in the yard, engaging in sports like playing ball, running, skipping rope, etc. On the other hand, adults should also incorporate these kinds of physical activities into their normal living with the WHO recommendation of having 150 minutes of physical activity training with moderate intensity [8]. In the end, the main problem lies in the fact that even after all these recommendations given to the children or the adults; only less than half of them involved themselves in any kind of physical activity (**Figure 1**) [9].

In 2018, Arto J. Pesola tried to incorporate different types of superpark activities among individuals. Both adults and children were included in the research that he conducted and they were made to play different kinds of super activities like hockey, baseball, street games, floorball skating, and enjoying themselves on trampoline. These certain activities were randomized to the participants for about 3 minutes given

Figure 1.
Children playing football *(Football encourages teamwork, social interaction, and physical fitness, fostering mental well-being in children by reducing stress and enhancing their social and emotional skills).*

the break period of 2 minutes in each. While performing these activities, their cardiovascular system such as heart rate along with their neuromuscular system stimulations were measured repeatedly. Initially, both the adults and the children were putting their minimum effort into all kinds of major activities. After a little push, they tried to perform at their maximum level by executing challenging, vigorous intensity activities. In the end, almost everyone contributed toward meeting the criteria of WHO recommendations [10].

1.2 Physical activity and women

Now, there is a little concern about physical activities and their awareness among women. Back in the day, even when recreational activities were found easily, women rarely engaged themselves in them. Most of them were not even familiar with that term. So, they rarely were found doing something like that [11]. They were considered to be more responsible toward household activities and were told repeatedly that they should be devoted to it completely like the chores of everyday living [12]. It was considered the norm regarding the standard customs of that time [13]. While both genders need to be able to have some fun during their busy lives, women were pushed to put more effort into their families rather than her selves. It can be found easily in the nineteenth century, and as it was the normal action, people hardly questioned any of it [14].

Interestingly, even when both were working just as much as the same while in their domains, men were appreciated more for their effort most of the time. The main problem was that even when doing some household work was not some kind of misfortune for them, getting no appreciation while being busy all day with having little to no time for themselves was one of the main trigger points [15]. One of the main reasons for women to feel fatigue, nervousness, and exhaustion, even in the early years of life is depression. It conveys that the depression was dominant even when it was not that known and accepted much at that time. So, depression was found commonly in some women without them being aware of it at all [16].

Just like that, even in today's time, women face many gender-specific barriers [17]. Women face unique biological and societal challenges that influence their physical activity levels. Life stages such as pregnancy, menopause, and post-menopause are associated with hormonal fluctuations that impact metabolism, bone density, and muscle strength [18]. Additionally, societal expectations, caregiving responsibilities, and limited access to resources create gender-specific barriers to physical activity. These challenges are particularly pronounced in low and middle-income countries, where cultural and socioeconomic factors further exacerbate health inequities [19]. These factors coupled with the rising prevalence of sedentary behaviors exacerbate health disparities, particularly in low and middle-income countries. However, evidence indicates that consistent physical activity can significantly reduce all-cause mortality by mitigating these kinds of risks [20].

Evidence Consistently supports the role of physical activity in reducing all-cause mortality among women. Regular exercise improves cardiovascular health, lowers blood pressure, enhances glucose metabolism, and supports weight management [21]. These benefits collectively reduce the risk of chronic illnesses that are leading contributors to women's mortality. Moreover, physical activity positively impacts mental health, reducing stress, anxiety, and depression which are often the overlooked contributors to overall health decline [22].

In the 1950s, social psychologists Irwin M. Rosenstock and his team proposed The Health Belief Model (HBM), a psychological framework to explain and predict health-related behaviors by focusing on individual beliefs and perceptions. It identifies factors that influence whether a person adopts preventive measures or engages in health-promoting activities. These factors include their perceived risks, benefits, barriers, and motivation to act. Applying the Health Belief Model provides a comprehensive framework to understand and promote the role of physical exercise in reducing mortality risk among women. By addressing each component—susceptibility, severity, benefits, barriers, cues to action, and self-efficacy—health interventions can be more effectively tailored to encourage women to engage in regular physical activity, thereby enhancing their overall health outcomes (**Figure 2**) [23].

1.3 Physiotherapist: A catalyst for change among women

Now as we have already talked about letting the women engage themselves in physical activities, professional help procuring some additional benefits can be a desirable addition. The help can be provided by many kinds of professionals, sports coaches, gym trainers, and lastly physiotherapists. Through strategic interventions that encompass aerobic, resistance, and functional training; physical therapists play a critical role in empowering women to avert premature demise while enhancing their quality of life [24]. For physical therapists, addressing this public health challenge involves not only promoting physical activity but also tailoring interventions to meet the unique physiological and psychological needs of women across various life stages [25].

Physical therapists are at the forefront of preventive health strategies. They design tailored physical activity programs that address women's unique needs while overcoming societal and physiological barriers. By incorporating aerobic, resistance, and functional training exercises into therapeutic regimens, physiotherapists enable

Figure 2.
Women indulging themselves in some healthy physical activities - designed using Canva *(This image shows the combination of controlled body movements that enhance core stability and posture, targeting the abdominal and lower back muscles essential for overall strength).*

women to enhance their overall health and avert premature demise. Their role extends beyond physical health, as they also address psychosocial well-being through education and empowerment, fostering long-term adherence to active lifestyles [26].

2. Influence physical activity holds over women

As we have already discussed physical activity can completely redefine someone's life. Engaging in regular physical activity significantly reduces the risk of chronic illnesses such as cardiovascular disease, diabetes, and osteoporosis, which are major contributors to women's mortality rate. Furthermore, it enhances overall mental resilience and promotes longevity by fostering a balanced and healthier lifestyle. To summarize it completely whether it is a physical change or a psychological change, exercise or physical activity holds more influence and value on women's lives than what can be described [27].

Some of the advantages it has on the women's body physically and mentally are given below:

Prevention of cardiovascular disease: Cardiovascular disease (CVD) has remained the leading cause of death in women globally, often presenting with atypical symptoms that complicate timely diagnosis. Physical activity serves as a potent non-pharmacological intervention by enhancing endothelial function, improving lipids profiles, and reducing blood pressure [28]. For instance, aerobic exercises like brisk walking or swimming can lower low-density lipoprotein (LDL) cholesterol while increasing high-density lipoprotein (HDL) cholesterol, reducing further the risk of coronary artery disease [25]. Furthermore, regular physical activity can decrease systemic inflammation, as evidenced by lower C-reactive protein levels, a critical marker of cardiovascular health [29, 30].

Enhancement of respiratory health: Chronic respiratory conditions such as asthma and chronic obstructive pulmonary disease (COPD) disproportionately affect women, especially in low and middle-income countries with high exposure to indoor pollutants. Physical activity, particularly breathing exercises and aerobic conditioning can strengthen the normal respiratory muscles, improving lung function and reducing dyspnea. This can further reduce the risk of complications and mortality associated with respiratory illnesses [31].

Combating osteoporosis and fall-related mortality: Postmenopausal women are particularly susceptible to osteoporosis, which significantly increases the risk of fractures and associated mortality. Resistance training and weight-bearing exercises stimulate bone remodeling and improve bone mineral density [32]. By enhancing muscle strength, balance, and proprioception; physical activity can not only prevent fractures but can also reduce the likelihood of falls, a major cause of mortality among older women. Physical therapists play a key role in designing individualized exercise programs that address bone health while minimizing injury risk (**Figure 3**) [33].

Reduction of cancer-related mortality: Physical Activity has been associated with the lower risks of breast and endometrial cancers, two of the most prevalent malignancies in women. Technically, exercise regulates insulin sensitivity, reduces circulating estrogen levels, and improves immune surveillance which collectively inhibits cancer progression. For breast cancer survivors, physical therapy-driven rehabilitation programs can improve functional capacity, alleviate treatment side effects, and further enhance psychological well-being, reducing overall mortality [34].

Mental health and stress resilience: Depression and anxiety are quite prevalent among women due to hormonal fluctuations and these psychosocial stressors are the significant predictors of premature mortality. Physical activity, particularly activities like dancing, cycling, or group aerobics can modulate stress responses through the hypothalamic-pituitary-adrenal axis and increase serotonin and dopamine levels. Improved mental health can not only extend life expectancy but can also enhance the adherence level of an individual to other health-promoting behaviors [35].

Mitigation of obesity and metabolic syndrome: Obesity, a key driver of non-communicable diseases, disproportionately affects women, particularly after menopause. Regular physical activity can improve insulin sensitivity, promoting fat oxidation and further enhancing metabolic flexibility. Physical therapists advocate for a combination of aerobic and resistance training to achieve sustainable weight loss and reduce the risk of complications such as diabetes and cardiovascular disease, both of which contribute significantly to mortality [36].

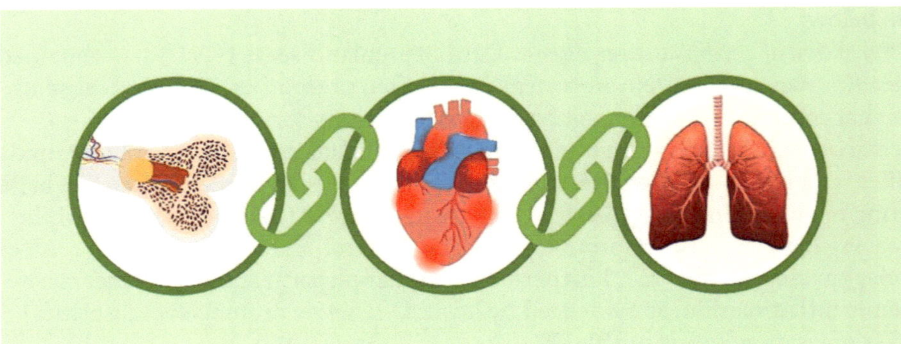

Figure 3.
Influence of physical activity on body organs - designed using Canva *(Visualizing the importance of physical activity on the normal organs in people. Interestingly, heart and lungs both are the main organs of the body).*

Improvement in immune function: Women's immune systems can be compromised due to stress, aging, or underlying health conditions increasing susceptibility to infections and chronic diseases. Regular moderate-intensity "physical activity" can enhance immune surveillance by boosting the natural killer (NK) cell activity, improving lymphocyte circulation, and reducing chronic inflammation. This immunomodulatory effect lowers the risk of infections and inflammatory diseases, thereby indirectly contributing to reduced mortality [37].

Regulation of hormonal imbalances: Physical activity profoundly influences hormonal health, particularly during transitional phases such as puberty, pregnancy, and menopause. For instance, regular exercise can modulate the levels of estrogen and progesterone, reducing the risks associated with hormonal imbalances like polycystic ovarian syndrome (PCOS) and hormone-driven cancers. During menopause, exercise alleviates symptoms such as hot flashes, mood swings, and sleep disturbances, while lowering the risk of chronic conditions such as osteoporosis and cardiovascular disease [38].

Alleviation of Pregnancy Related Complications: Pregnancy complications such as gestational diabetes, pre-eclampsia, and excessive weight gain contribute significantly to maternal mortality. Tailored physical activity programs during pregnancy such as prenatal yoga or aquatic exercises can improve maternal cardiovascular fitness, regulate blood pressure, and enhance glucose metabolism. Furthermore, exercise can reduce the risk of preterm birth and cesarean delivery while improving postpartum recovery, thereby safeguarding maternal health [39].

Reduction of stroke incidence and mortality: Stroke is a leading cause of disability and mortality among women, with the risks increasing during post menopause due to hormonal and vascular changes. Regular physical activity can improve cerebral blood flow, reducing hypertension and mitigating risk factors such as high cholesterol

Health Area	**Physical activity that will help**	**Benefits of applied physical activity**
Cardiovascular Health	Aerobic programs (e.g., brisk walking and swimming) are tailored to cardiovascular fitness levels.	Reduces blood pressure, improves cholesterol levels, and decreases systemic inflammation [29].
Bone Health	Weight-bearing and resistance exercises to enhance balance and prevent fractures.	Increases bone mineral density, reduces osteoporosis risk, and lowers fall-related mortality [33].
Cancer Prevention	A complete rehabilitation plans for cancer survivors to improve functional capacity and reduce treatment effects.	Regulates insulin sensitivity, lowers estrogen levels, and boosts immune surveillance [34].
Mental Health	Group activities or stress-relieving exercises like yoga to improve emotional well-being.	Improves moods, reduces anxiety and depression, and enhances resilience to stress [35].
Metabolic Health	Combining aerobic and resistance exercises is recommended to target metabolic syndrome and obesity.	Enhances insulin sensitivity, promotes fat oxidation, and prevents obesity-related complications [36].
Longevity and Quality of Life	Encourage consistent physical activity for sustainable health improvements and better quality of life.	Combats chronic diseases enhances physical fitness, and improves life expectancy [36].

Table 1.
Summarization of some physical activities/exercises along with their benefits for women in the prevention of certain health conditions.

and obesity. Studies show that women engaging in moderate aerobic exercise, such as cycling or jogging, experience a significantly lower risk of ischemic stroke. Post-stroke rehabilitation programs led by physical therapists can enhance functional recovery and reduce secondary complications, ensuring better survival outcomes [40].

Promotion of longevity and quality of life in aging women: Physical activity can directly impact longevity by reducing all-cause mortality rates. In older women, regular exercise can preserve functional independence, preventing cognitive decline and enhancing the quality of life. Activities like tai chi and brisk walking can improve cardiovascular fitness, muscle strength, and balance, reducing the risk of falls and frailty. Additionally, these activities promote social interaction and mental stimulation further contributing to well-being [41].

Physical Activity is not merely a lifestyle choice for women; it is a critical intervention for averting premature demise. By addressing barriers to activity, empowering women with knowledge, and leveraging community support, physical activity can help bridge the gap between inactivity and vibrant health significantly reducing mortality rates and enhancing overall well-being (**Table 1**).

This table will offer some information regarding the type of exercise that can help in preventing or improving certain health conditions. A complete guide to the benefits associated with the exercises can provide further guidance for the people.

3. Categories of physical activities for women

There are multiple kinds of physical activities a woman can adopt to make changes in their life, each having their benefits. It depends on the goal of a woman to choose a specific physical activity. The effects of these activities vary largely based on their age, previous lifestyle, current lifestyle, consistency, motivation of the person, health issues, and most importantly diet, which has a direct link with the effects of physical activity [42].

Here is an overview of different types of exercises for women to choose from along with examples:

Aerobic exercises: These are the activities that can enhance the heart rate and breathing for a sustained period, improving the efficiency of the cardiovascular and respiratory systems. Running, cycling, swimming, brisk walking, and dancing can be quite helpful for women as they are aerobic exercises [42].

Strength training: Exercises that use resistance to induce muscle contraction, improving muscle strength, endurance, and size. Examples are weightlifting, resistance band exercises, and bodyweight exercises like push-ups and squats. Nowadays, women are trying to involve themselves more in this kind of exercise to improve their muscular capability [43].

Flexibility exercises: Movements that improve the range of motion in muscles and joints, reducing stiffness and preventing injuries. Such as stretching routines, yoga poses, dynamic stretches, and static stretches. Women like to involve themselves in these types of activities to improve their elasticity specifically ballet dancers and gymnastics (**Figures 4** and 5) [44].

Balance exercises: Activities that improve the ability to maintain the body's center of gravity over its base of support. Some examples are standing on one foot, balance board exercises, yoga, tai chi, and tandem walk. Normally, balance exercises can be quite beneficial for women of any age but as the focus of the balance exercises is the improvement in stability, these exercises can be really helpful for older age adult women [45].

Figure 4.
Some stretching exercise poses *(A toe touch stretch that helps to enhance hamstring flexibility, improve spinal mobility, and relieve tension in the lower back, promoting overall body relaxation).*

Figure 5.
Some stretching exercise poses *(Pigeon pose that is a deep hip opener helps in stretching the hip flexors, glutes, and lower back, enhancing flexibility and alleviating tension in the pelvic region).*

Figure 6.
Plank for core stability *(Plank strengthens the core muscles when holding this certain position, it can also improve coordination, flexibility and balance).*

Core stability exercises: Exercises targeting the deep muscles to stabilize and support the spine and pelvis. Like planks, side bridges, bird dogs, and Pilates. Just like flexibility exercises, core stability exercises are also quite common among women (**Figure 6**) [46].

Figure 7.
Yoga pose *(This stretching technique focuses on flexibility and balance, emphasizing the joint mobility and reduce muscle stiffness. It improves posture as well as relaxes all the flexor muscles of the body).*

High-intensity interval training (HIIT): A form of exercise which involves short bursts of intense activity alternated with periods of rest or lower-intensity exercise. Short bursts of intense exercises like sprinting, burpees, or jump squats alternated with rest are the common examples of this training regimen. This kind of exercise plan is becoming common among women of today's time, specifically those who cannot indulge themselves completely in strength training exercises [47].

Mind-body exercises: To enhance overall well-being, mind-body exercises are considered to improve mental concentration, physical movement, and steady breathing. They can improve flexibility, strength, and balance. Yoga, Pilates, and tai chi can help in relaxation. Another common example of activity found among women is to improve their mental health (**Figures 7** and **8**) [48].

Figure 8.
Yoga pose *(Lunges are a dynamic lower-body exercise that targets the quadriceps, hamstrings, and glutes, improving leg strength, balance and flexibility).*

4. General physical activity recommendations for women

Activities should align with the participant's abilities, emphasizing safety and enjoyment. General physical activity guidelines vary according to the age and demand of the person. Adapting some physical activity ensures better health and enhanced benefits for all age groups. Two main things need to be focused on while customizing a physical activity program for a woman.

Considering exercise intensity: Moderate-intensity activities can be beneficial for most women, while higher-intensity activities may be suitable for those with higher fitness levels. For example, a minimum of 3–5 days/week, depending on the individual's physical capacity and sports, starting with low to moderate intensity, gradually progressing to higher levels as tolerated.

Consulting a healthcare provider: It is important to consult with a healthcare provider before starting a new exercise program, especially for women with underlying health conditions [49].

The average amount of physical activity required by the women in different age groups is as follows (**Figure 9**).

5. How treatment plans of women differ from the men

Several factors can clearly state the difference between how a treatment plan is designed for men and women. Those factors will be discussed here as how they can lead to an understandable distinction between the two genders. It further depends on the goals that are found to be different in both genders. These factors are widely documented in various studies.

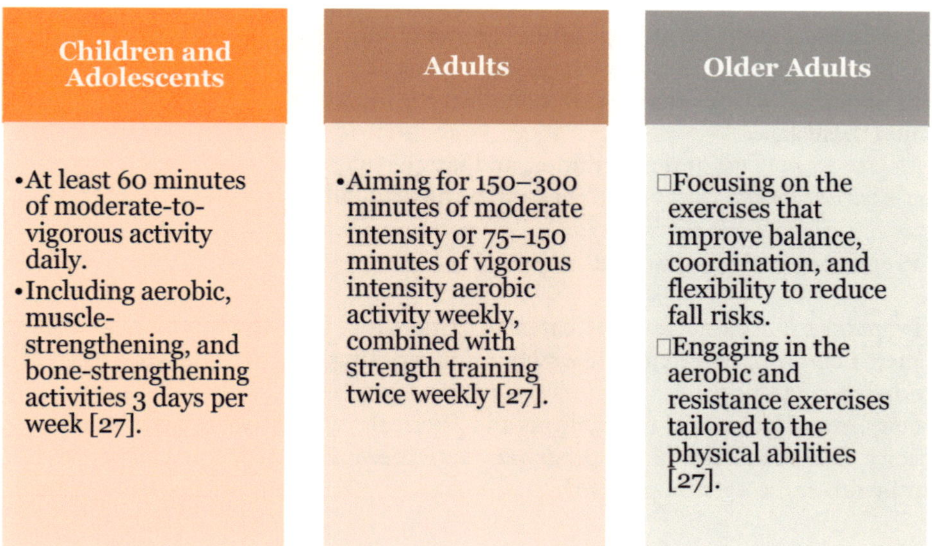

Figure 9.
Physical Activity recommendations across different age groups. (This figure outlines exercise guidelines for children, adults, and older adults, emphasizing aerobic, strength, and balance training designed for each age group's needs.)

Participation levels: Men are generally more likely to engage in vigorous physical activities, while women often participate in moderate activities. A study among university students showed that men had higher overall physical activity levels than women. Women tended to have lower intrinsic motivation, relying more on external factors such as social encouragement or perceived health benefits to participate in activities [42].

Motivational factors: Men are typically motivated by competition and performance improvement, whereas women are more likely to engage in physical activity for health benefits, stress relief, and social interaction. Women also report higher levels of external motivation and demotivation compared to men [43].

Barriers to participation: Common barriers for women include lack of time, cultural norms, and safety concerns while men may face fewer societal restrictions but can also experience pressure to perform at high levels in competitive environments [43].

Implications for public health: Gender-sensitive approaches are essential for promoting physical activity. This includes designing programs that address the unique needs and motivations of men and women, creating safe environments for all genders, and addressing societal norms that discourage female participation in certain sports or activities [42].

Hormonal influences: It is one of the major gender differences that affect the physical and functional gains in women. Programs should consider gender-specific physiological responses to optimize performance and reduce injury risks. Training and recovery plans for women can benefit from aligning with menstrual cycle phases [43].

Hormonal fluctuations throughout a woman's life significantly influence physical activity performance, injury risk, and overall well-being. Let us delve into the key stages, their implications, and the type of activity that can provide some help during this stage:

5.1 Menstrual cycle

Follicular phase: Rise in the estrogen levels leading to increased energy and improved mood. This is a great time for higher-intensity workouts like HIIT or strength training.

Ovulatory phase: Progesterone levels rise, potentially causing slight fatigue and joint stiffness. Lower-impact activities like swimming or yoga will be beneficial.

Luteal phase: Progesterone levels peak, leading to potential mood swings and decreased energy. Considering lower-intensity workouts or rest days are best for women these days.

Menstruation: Discomfort, cramps, and fatigue may affect exercise tolerance. Gentle activities like yoga or walking will be helpful [44].

5.2 Pregnancy and postpartum

Pregnancy: Hormonal changes can lead to increased fatigue, joint pain, and altered balance. Low-impact activities like swimming, walking, and prenatal yoga are recommended during this time.

Postpartum: Hormonal fluctuations and physical recovery can impact exercise capacity. Gradual return to pre-pregnancy activities is essential, considering pelvic floor health and core strength [44].

5.3 Menopause

Perimenopause and menopause: Declining estrogen levels can lead to decreased bone density, muscle mass, and increased risk of cardiovascular disease. Weight-bearing exercises, resistance training, and regular physical activity are crucial [44].

6. Engaging women in fun physical activities

Till now, we have talked about the idea of women's involvement in physical activities leading to their better physical and psychological health. Then the type of activities along with its influence on the women having better health during every stage of their lives. Even the idea of how a better treatment plan can be made according to certain health conditions or just for the improvement of the well-being of the women was discussed above. Nevertheless, the idea of just forcing women into some random physical activity was not enough. The other thing needed was to be able to have some fun while exercising. It is also a source of improving one's quality of life. It is a fact that poking fun at doing recreational activities or performing healthy activities is not some kind of modern idea. This concept is almost as old as about more than a century ago, but it only garners more attention cause modern life has found many kinds of leisure activities than the ones found back in the day. Its strength is increasing in different areas of society [50].

Like WHO has stated in one of its other studies that,

> *"If a person whether a man or a woman wants to increase their quality of life there are two factors that need improvement; one is having a habit of healthy eating and second one is involving themselves in recreational activities" (***Figure 10***) [51].*

For this reason, doing just physical activity without any goal or motivation is nothing but just some extra work so it is better to have some kind of recreational and fun activities. These physical activities would not only bring positive effects on the physical body but also procure some healthy mental effects especially. It is

Figure 10.
Girl doing bicycling (recreational activity) *(Bicycling is an excellent social activity that promotes cardiovascular health, builds endurance, and boosts mood, helping women and girls stay physically active while enjoying the outdoors).*

particularly helpful in dealing with the depression that is being discussed before [52]. Unfortunately, in today's time, even though women have obtained more free time for themselves than back in the day, some of them still try to ignore their physical and mental health by neglecting themselves [53]. A logical and sensible reasoning is that when people have some free time, they should try to engage themselves in some kind of fun activity occasionally that will bring them more happiness and self-satisfaction in life. Gradually, when they find these activities enjoyable, leading to procuring some positive effects resulting in having a healthy body with a healthy mind, they will try to incorporate it into their everyday lives. So, it can be said that having fun while performing physical activity can lead to positive mental health, subsiding depression [54].

Seyda Alanoglu tried to figure out the effects of recreational activities on women's happiness levels. For this purpose, he conducted research in 2020 regarding the effects of regularly doing recreational and fun activities on adult women's happiness, stress, and satisfaction levels. It was a 10-week study in which 199 adult women volunteered to take part. The results indicated that adult women must be more active in their lives if they wish to have a healthy life. If they take part in some regularly performed recreational and fun activities, they can easily get rid of the buildup of stress and fatigue levels from their lives. It can easily subside the exhaustion and, subsequently, the depression from their lives [55].

7. Professional sports: A stressful event for women

As already been discussed how the concept of women engaging themselves in some physical activities has gained more importance in recent years. One more noticeable thing is their involvement in professional sports. Being able to participate in professional sports is another way of putting someone under immense pressure, deteriorating their mental health. It does not matter whether the person suffering from it is a male or a female; the concept of winning is enough to create enough burden, but due to all the reasons discussed above it is becoming somewhat extra burdensome for women [56].

Some studies have shown that the biggest reason a professional female player feels that much pressure is not because of the chance of losing but because of the involvement of coaches who do not pay much attention to the players. As women are still in the process of gaining recognition in some sports, further pressure just pushes them away without giving them any chance to even participate. Participants should be able to feel happy and being able to feel proud of themselves rather than ashamed of themselves during any adversity. With the increasing pressure on the professional players, specifically on the women, due to some complex external factors, they started feeling inferior to others [57].

A study was conducted in 1999 by Karla Henderson to figure out how fun can be incorporated into professional physical activities for women like sports so that the participants will be able to enjoy the games they play rather than feeling stressed all the time. She showed her concerns of putting a tremendous amount of pressure on the women athletes or participants involving games, leading to the increasing level of anxiety and depression in them that can further deteriorate their normal mental health as well as physical health. It is important to lead recreational professionals' coaches to make sure to motivate their athletes positively rather than downsizing them through negative comments so that they would not feel neglected [58].

8. Community-based activities: A breakthrough

Why performing individual physical activities and leisure activities or some kind of sports can be beneficial, the community-based activities are proven to be more effective than them [59]. Different studies have indicated that those females who involve themselves in community-based activities tend to enjoy themselves more than when they perform these activities alone. The reason was having some kind of emotional support from the other group members [60].

In group activities, they can enjoy themselves completely leading to more kinds of psychological benefits. Different kinds of sports activities and programs are structured to develop some positive outcomes within women, for example, a positive youth development (PYD) approach [61]. This approach was designed to improve the psychosocial development of the females like a group of females involving themselves toward improving their physical and mental health. This will not only affect their physical body leading to better physical health while also developing their mental health cognition and their improving social interactions [62].

There is also new research that was conducted in 2015 by Corliss Bean titled "Girls Just Want to Have Fun" basically concerning the effective strategies and their outcomes of performing physical activities in everyday living among females. It was a community-based program about the physical activity of the female youth and developing some life skills for them. Two categories of females were recruited in this study, one female participant and the second the program leaders. The results of this study concluded that both the youth participants, as well as the leaders, perceived this program as incorporating effective strategies and generating some positive outcomes [63].

9. Conclusion

At the end of the study, it can be concluded that physical activity can easily change the lives of women for the better. A small and simple effort toward their health can completely reshape them, just like explained in the 1950s Health Belief Model. There are different categories of physical activities, such as exercise regimens, sports, or recreational activity. A customized physical therapy plan cannot only improve their overall well-being but can also combat different kinds of women-related health conditions. Different professionals help in customizing these physical activity regimens according to the needs and goals of the women. Still, simply doing physical activity helps but involving fun in them can be quite innovative leading to some additional benefits for women. It also includes the involvement of community-based activity programs for women. All of them have a significant impact on the mental health of the women along with the improvement of their physical health. It also helps women build some social connections. Altogether, physical activity is just essential for women in every aspect of life.

Acknowledgements

This modest and humble effort is purely dedicated to the uplifting tenderness and love of our *parents*. This thank you goes to the parents of all involved authors who are always praying for the success and good health of their children.

Conflict of interest

The authors declare no conflict of interest.

Abbreviations

CVD	cardiovascular disease
COPD	chronic pulmonary obstructive disease
HBM	health belief model
HIIT	high interval intensity training
HDL	high-density lipoprotein
LDL	low-density lipoprotein
PCOS	polycystic ovarian syndrome
PYD	positive youth development
WHO	World Health Organization

Author details

Hifza Ahmed[1*], Haleema Sadia[2], Ulvina Riaz[3] and Irtasam Ahmad[4]

1 Fatima Zia Medical Complex, Faisalabad, Pakistan

2 Faisalabad Teaching Hospital, Faisalabad, Pakistan

3 Bright Hope Foundation, Faisalabad, Pakistan

4 Madina Teaching Hospital, Faisalabad, Pakistan

*Address all correspondence to: hifzaahmed77@gmail.com

References

[1] Herrmann SD, Willis EA, Ainsworth BE, Barreira TV, Hastert M, Kracht CL, et al. 2024 adult compendium of physical activities: A third update of the energy costs of human activities. Journal of Sport and Health Science. 2024;**13**(1):6-12

[2] Shibasaki H. Cortical activities associated with voluntary movements and involuntary movements. Clinical Neurophysiology. 2012;**123**(2):229-243

[3] Hills AP, Mokhtar N, Byrne NM. Assessment of Physical Activity and Energy Expenditure: An Overview of Objective Measures. Frontiers in Nutrition. 2014;**1**

[4] Garland T Jr, Schutz H, Chappell MA, Keeney BK, Meek TH, Copes LE, et al. The biological control of voluntary exercise, spontaneous physical activity and daily energy expenditure in relation to obesity: Human and rodent perspectives. Journal of Experimental Biology. 2011;**214**(2):206-229

[5] Benjamin RM. The surgeon general's vision for a healthy and fit nation. Public Health Reports®. 2010;**125**(4):514-515

[6] Haapala EA. Cardiorespiratory fitness and motor skills in relation to cognition and academic performance in children - A review. Journal of Human Kinetics. 2013;**36**:55-68

[7] Biddle SJ, Asare M. Physical activity and mental health in children and adolescents: A review of reviews. British Journal of Sports Medicine. 2011;**45**(11):886-895

[8] WHO Guidelines Approved by the Guidelines Review Committee. Global Recommendations on Physical Activity for Health. Geneva: World Health Organization; 2010

[9] Tammelin TH, Aira A, Hakamäki M, Husu P, Kallio J, Kokko S, et al. Results from Finland's 2016 report card on physical activity for children and youth. Journal of Physical Activity and Health. 2016;**13**(11 Suppl 2):S157-Ss64

[10] Liikanen V, Pesola A. Physical Fun: Exercise, Social Relations and Learning in SuperPark. Grano: Finland University of Applied Sciences 2018. ISBN: 978-952-344-076-0 (nid.) ISBN: 978-952-344-077-7 (PDF) ISSN: 2489-2459 ISSN: 2489-4575 (ebook)

[11] Koca C. Gender relations in physical education and sport. Hacettepe Journal of Sport Sciences. 2006;**17**(2):81-99

[12] Powell GN, Greenhaus JH. Sex, gender, and decisions at the family→work interface. Journal of Management. 2010;**36**(4):1011-1039

[13] Bourdieu P. Sport and social class. Social Science Information. 1978;**17**(6):819-840

[14] Budak G, Doğan HZ, Harlak AGH. Çalışan kadınların sorunları: bir toplumsal değişme araştırması. Sosyal Politika Çalışmaları Dergisi. 1991;**1**(1). ISSN: 1303-0256

[15] Aydoğan İ. Kadın öğretim elemanlarının boş zamanlarını değerlendirme ekinlikleri. Erciyes Üniversitesi Sosyal Bilimler Enstitüsü Dergisi. 2006;**1**(21):217-232

[16] Cordes KA. Applications in Recreation & Leisure: For Today and the Future. USA: Sagamore-Venture Publishing; 2013. ISBN/ISSN: 978-1-57167-700-6

[17] White PR. The Menopausal Self: Identity Construction and Gender Norms in Menopause Apps: A Thesis Presented in Partial Fulfilment of the Requirements for the Degree of Master of Arts in Psychology at Massey University, Palmerston North. New Zealand: Massey University; 2024

[18] Kodete CS, Thuraka B, Pasupuleti V, Malisetty S. Hormonal influences on skeletal muscle function in women across life stages: A systematic review. Muscles. San Francisco, CA: Jossey-Bass A Wiley Brand; 2024;**3**(3):271-286

[19] Peng B, Ng JY, Ha AS. Barriers and facilitators to physical activity for young adult women: A systematic review and thematic synthesis of qualitative literature. International Journal of Behavioral Nutrition and Physical Activity. 2023;**20**(1):23

[20] Golabi P, Gerber L, Paik JM, Deshpande R, de Avila L, Younossi ZM. Contribution of sarcopenia and physical inactivity to mortality in people with non-alcoholic fatty liver disease. JHEP Reports. 2020;**2**(6):100171

[21] Samuel PO, Edo GI, Emakpor OL, Oloni GO, Ezekiel GO, Essaghah AEA, et al. Lifestyle modifications for preventing and managing cardiovascular diseases. Sport Sciences for Health. 2024;**20**(1):23-36

[22] Belvederi Murri M, Folesani F, Zerbinati L, Nanni MG, Ounalli H, Caruso R, et al. Physical activity promotes health and reduces cardiovascular mortality in depressed populations: A literature overview. International Journal of Environmental Research and Public Health. 2020;**17**(15):5545

[23] Glanz K, Rimer BK, Viswanath K. Health Behavior: Theory, Research, and Practice. John Wiley & Sons; 2015

[24] Coppi F, Bucciarelli V, Solodka K, Selleri V, Zanini G, Pinti M, et al. The impact of stress and social determinants on diet in cardiovascular prevention in young women. Nutrients. 2024;**16**(7):1044

[25] Lopez-Jimenez F, Almahmeed W, Bays H, Cuevas A, Di Angelantonio E, le Roux CW, et al. Obesity and cardiovascular disease: Mechanistic insights and management strategies. A joint position paper by the world heart federation and world obesity federation. European Journal of Preventive Cardiology. 2022;**29**(17):2218-2237

[26] Lucini D, Pagani M. Exercise prescription to foster health and well-being: A behavioral approach to transform barriers into opportunities. International Journal of Environmental Research and Public Health. 2021;**18**(3):968

[27] Zábó V, Csiszar A, Ungvari Z, Purebl G. Psychological resilience and competence: Key promoters of successful aging and flourishing in late life. GeroScience. 2023;**45**(5):3045-3058

[28] Tjahjono CT, Pramudya A. Comprehensive management of hypertension: Enhancing non-pharmacological treatments. Medical Research Archives. 2023;**11**(3). ISSN 2375-1924

[29] Ungvari Z, Fazekas-Pongor V, Csiszar A, Kunutsor SK. The multifaceted benefits of walking for healthy aging: From blue zones to molecular mechanisms. Geroscience. 2023;**45**(6):3211-3239

[30] Van't Klooster C, van Der Graaf Y, Ridker P, Westerink J, Hjortnaes J, Sluijs I, et al. The relation between healthy lifestyle changes and decrease in systemic inflammation in patients

with stable cardiovascular disease. Atherosclerosis. 2020;**301**:37-43

[31] Boutros P, Kassem N, Boudo V, Sié A, Munga S, Maggioni MA, et al. Understanding the risk factors, burden, and interventions for chronic respiratory diseases in low-and middle-income countries: A scoping review. Public Health Reviews. 2024;**45**:1607339

[32] Kim S-W, Seo M-W, Jung H-C, Song J-K. Effects of high-impact weight-bearing exercise on bone mineral density and bone metabolism in middle-aged premenopausal women: A randomized controlled trial. Applied Sciences. 2021;**11**(2):846

[33] Stevenson J, Society MACotBM. Prevention and treatment of osteoporosis in women. Post Reproductive Health. 2023;**29**(1):11-14

[34] Larson EA, Dalamaga M, Magkos F. The role of exercise in obesity-related cancers: Current evidence and biological mechanisms. In: Seminars in Cancer Biology. Amsterdam, Netherlands: Elsevier Ltd; 2023

[35] Bendau A, Petzold MB, Kaminski J, Plag J. Exercise as treatment for" stress-related" mental disorders. Current Neuropharmacology. 2024;**22**(3):420-436

[36] Al-Mhanna SB, Rocha-Rodriguesc S, Mohamed M, Batrakoulis A, Aldhahi MI, Afolabi HA, et al. Effects of combined aerobic exercise and diet on cardiometabolic health in patients with obesity and type 2 diabetes: A systematic review and meta-analysis. BMC Sports Science, Medicine and Rehabilitation. 2023;**15**(1):165

[37] Strzelec M, Detka J, Mieszczak P, Sobocińska MK, Majka M. Immunomodulation—A general review of the current state-of-the-art and new therapeutic strategies for targeting the immune system. Frontiers in Immunology. 2023;**14**:1127704

[38] Bagga SS, Tayade S, Lohiya N, Tyagi A, Chauhan A. Menopause dynamics: From symptoms to quality of life, unraveling the complexities of the hormonal shift. Multidisciplinary Reviews. 2025;**8**(2):2025057

[39] Childs C, Sandy-Hodgetts K, Broad C, Cooper R, Manresa M, Verdú-Soriano J. Birth-related wounds: Risk, prevention and management of complications after vaginal and caesarean section birth. Journal of Wound Care. 2020;**29**(Sup11a):S1-S48

[40] Li X, He Y, Wang D, Rezaei MJ. Stroke rehabilitation: From diagnosis to therapy. Frontiers in Neurology. 2024;**15**:1402729

[41] Bar-Tur L. Fostering well-being in the elderly: Translating theories on positive aging to practical approaches. Frontiers in Medicine. 2021;**8**:517226

[42] Olson RD, Vaux-Bjerke A, Quam JB, Piercy KL, Troiano RP, George SM, et al. Physical activity guidelines for Americans. NADAR! SWIMMING MAGAZINE-Periódico científico em esportes e fitness aquático-natação, pólo aquático, nado sincronizado, saltos ornamentais, travessias aquáticas. SWIMMING MAGAZINE; 2023. ISSN 2764-4677

[43] Kraemer WJ, Koziris LP, Ratamess NA, Häkkinen K, Triplett-McBride NT, Fry AC, et al. Detraining produces minimal changes in physical performance and hormonal variables in recreationally strength-trained men. The Journal of Strength and Conditioning Research. 2002;**16**(3):373-382

[44] Swain DP, Brawner CA, Medicine ACoS. ACSM's Resource Manual for Guidelines for Exercise Testing and Prescription. Philadelphia, Pennsylvania: Wolters Kluwer Health/Lippincott Williams & Wilkins; 2014. ISBN 9781609139568

[45] Shumway-Cook A. Motor Control: Translating Research into Clinical Practice. Philadelphia, Pennsylvania: Lippincoot Williams & Wilkins; 2007

[46] Akuthota V, Ferreiro A, Moore T, Fredericson M. Core stability exercise principles. Current Sports Medicine Reports. 2008;**7**(1):39-44

[47] Burgomaster KA, Howarth KR, Phillips SM, Rakobowchuk M, MacDonald MJ, McGee SL, et al. Similar metabolic adaptations during exercise after low volume sprint interval and traditional endurance training in humans. The Journal of physiology. 2008;**586**(1):151-160

[48] Cramer AO, Borsboom D, Aggen SH, Kendler KS. The pathoplasticity of dysphoric episodes: Differential impact of stressful life events on the pattern of depressive symptom inter-correlations. Psychological Medicine. 2012;**42**(5):957-965

[49] Weldon A, Duncan MJ, Turner A, LaPlaca D, Sampaio J, Christie CJ. Practices of strength and conditioning coaches: A snapshot from different sports, countries, and expertise levels. The Journal of Strength and Conditioning Research. 2022;**36**(5):1335-1344

[50] Ayhan C, Ekinci NE, Yalçın İ, Yiğit Ş. Investigation of constraints that occur during participation in leisure activities by high school students: A sample of Turkey. Education Sciences. 2018;**8**(2):86

[51] Soyer F, Yıldız NO, Demirel DH, Serdar E, Demirel M, Ayhan C, et al. The investigation of the relationship between the factors that prevent university students from attending to the recreational activities and the life satisfaction of the participants Üniversite öğrencilerinin rekreatif etkinliklere katılımlarına engel teşkil eden faktörler ile katılımcıların yaşam doyumları arasındaki ilişkinin incelenmesi. Journal of Human Sciences. 2017;**14**(2):2035-2046

[52] Tütüncü Ö. Rekreasyon ve rekreasyon terapisinin yaşam kalitesindeki rolü. Anatolia: Turizm Araştırmaları Dergisi. 2012;**23**(2):248-252

[53] Bedir F. Boş zaman (Rekreasyonel) aktivitelerinin stresin üstesinden gelebilme rolünün incelenmesi. Yayınlanmış Yüksek Lisans Tezi Sosyal Bilimler Enstitüsü Atatürk Üniversitesi Erzurum. Mediterranean Journal of Sport Science (MJSS) is licensed under a Creative Commons Attribution 4.0 International License CC BY-NC 4.0. Turkey: Hasan ŞAHAN; 2016

[54] Flow CM. The psychology of optimal experience. New York: HarperPerennial; 1990. pp. 303-336

[55] Alanoğlu S, Işık Ö, Ayhan C. The effect of regular recreational activities on adult women's stress, happiness, and life satisfaction levels. Progress in Nutrition. 2020;**22**(2)

[56] Howe CZ, Rancourt AM. The importance of definitions of selected concepts for leisure inquiry. Leisure Sciences. 1990;**12**(4):395-406

[57] Shaw SM, Bonen A, McCabe JF. Do more constraints mean less leisure? Examining the relationship between constraints and participation. Journal of leisure Research. 1991;**23**(4):286-300

[58] Henderson K, Glancy M, Little S. Putting the fun into physical activity. Journal of Physical Education, Recreation and Dance. 1999;**70**(8):43-45

[59] Coalter F. A wider Social Role for Sport: Who's Keeping the Score? Routledge. London: Routledge; 2007

[60] Gray CE, Larouche R, Barnes JD, Colley RC, Cowie Bonne J, Arthur M, et al. Are we driving our kids to unhealthy habits? Results of the active healthy kids Canada 2013 report card on physical activity for children and youth. International Journal of Environmental Research and Public Health. 2014;**11**(6):6009-6020

[61] Forneris T, Bean CN, Snowden M, Fortier M. Using youth-driven programs to encourage physical activity in adolescent girls: A preliminary study. Revue phénEPS/PHEnex Journal. 2013;**5**(1)

[62] Janssen I. Physical activity guidelines for children and youth. Applied Physiology, Nutrition, and Metabolism. 2007;**32**(S2E):S109-S121

[63] Bean C, Forneris T, Fortier M. Girls just wanna have fun: Understanding perceptions of effective strategies and outcomes in a female youth-driven physical activity-based life skills programme. Journal of Sport for Development. 2015;**3**(4):28-40

Section 3

Therapeutic Interventions, Mechanisms, and Clinical Applications

Chapter 9

Healing Waters: Aquatherapy for Mental Resilience and Injury Prevention

Aryan Shukla

Abstract

This chapter, Healing Waters: Aquatherapy for Mental Resilience and Injury Prevention, delves into the transformative therapeutic benefits of hydrotherapy in enhancing both physical and mental well-being. The unique properties of water—buoyancy, resistance, and temperature modulation—create an environment conducive to low impact and efficient movement making it ideal for injury prevention and rehabilitation. Hydrotherapy plays a pivotal role in alleviating physical strain, promoting joint health, enhancing muscular strength and flexibility without the stress associated with weight-bearing exercises. Beyond its physical advantages, water-based activities are shown to have profound psychological benefits. Hydrotherapy reduces stress, elevates mood, and fosters mental resilience by stimulating the parasympathetic nervous system, which promotes relaxation and emotional balance. The chapter highlights the strong interplay between mental health and physical recovery, showcasing how aquatic environments serve as a bridge to holistic healing. Furthermore, this chapter offers evidence-based insights into designing targeted aquatic exercise programmes tailored to diverse populations, such as athletes aiming to prevent injuries, individuals recovering from orthopaedic conditions, and those seeking enhanced mental health. By integrating practical applications and real-world examples, it underscores the potential of hydrotherapy to revolutionise traditional approaches to fitness, therapy, and mental healthcare, providing a holistic pathway to resilience and recovery.

Keywords: injury prevention, low-impact exercises, hydrotherapy, mental fitness, mental health

1. Introduction

Aquatic exercise, often referred to as hydrotherapy or water-based exercise, has been practised for centuries as a method of healing to enhance physical well-being. From ancient civilisations to modern rehabilitation, the unique properties of water have been harnessed for their therapeutic benefits. The therapeutic use of water can be traced back to ancient cultures. The Greeks and Romans made use of warm baths and thermal springs to restore physical vitality and alleviate mental stress. Roman baths, equipped with pools of varying temperatures, were central to public health and community well-being. In the East, traditional Chinese medicine incorporated water-based therapies to promote balance and energy flow in the body. During the nineteenth century, the development of hydrotherapy in Europe, pioneered by figures like Sebastian Kneipp, laid the foundation for modern aquatic therapy, emphasising water's curative potential for ailments such as arthritis and chronic pain [1]. Water possesses distinctive physical properties that make it an ideal medium for exercise and therapy. Buoyancy reduces the effects of gravity, allowing for low-impact movement that minimises stress on joints and muscles [2]. Hydrostatic pressure exerts a gentle compressive force on the body, promoting circulation and reducing swelling in injured tissues. Additionally, water's viscosity provides natural resistance, enabling controlled strength training and stability exercises without the need for heavy equipment. These characteristics make aquatic exercise accessible and beneficial for individuals of all fitness levels and health conditions. The therapeutic effects of water are deeply rooted in both physiology and psychology. Immersion in water triggers the parasympathetic nervous system, reducing stress and promoting relaxation. Studies have shown that warm water immersion increases blood flow to muscles, accelerating recovery and reducing muscle tightness [3].

Moreover, the buoyancy provided by water facilitates movement for individuals with limited mobility, enabling rehabilitation without exacerbating pain or injury. On a psychological level, the soothing nature of water fosters mindfulness and emotional well-being. Research has highlighted the positive impact of aquatic environment on mental health, including reduction in anxiety and depression [4]. The combination of physical support and sensory stimulation in water creates a unique environment conducive to healing both body and mind.

2. Mental resilience through aquatic exercises

Immersion in water has profound effects on the nervous system, promoting relaxation and emotional stability. Warm water stimulates the parasympathetic nervous system, leading to the release of neurotransmitters like serotonin, dopamine, and endorphins, which play key roles in enhancing mood and reducing anxiety [1]. The rhythmic movements involved in aquatic exercises further boost the release of these "feel-good" hormones, creating a natural calming effect on the brain (**Figure 1**). Additionally, hydrostatic pressure increases blood flow to the brain, improving oxygen and nutrient delivery, which is crucial for cognitive function and emotional balance [3]. Regular participation in aquatic activities has been shown to lower cortisol levels, a stress-related hormone, fostering long-term mental well-being. Aquatic exercises encourage mindfulness and self-awareness, essential components for the mind-body connection. The unique sensory environment of water enhances proprioception (awareness of body position), allowing participants to focus more deeply on their movements and breathing. Incorporating strategies like synchronised breathing with movements, gentle stretching, and mindful meditation in water amplifies this connection [4]. For example, yoga and Tai Chi-inspired movements adapted for water provide opportunities to integrate physical activity with mental focus. The resistance of water slows down movements, promoting intentionality and fostering a greater sense of control over the body and mind. The tranquil nature of water provides an ideal environment for enhancing focus and emotional regulation. The soothing effects of immersion reduce external distractions, allowing individuals to concentrate fully on their movements and mental

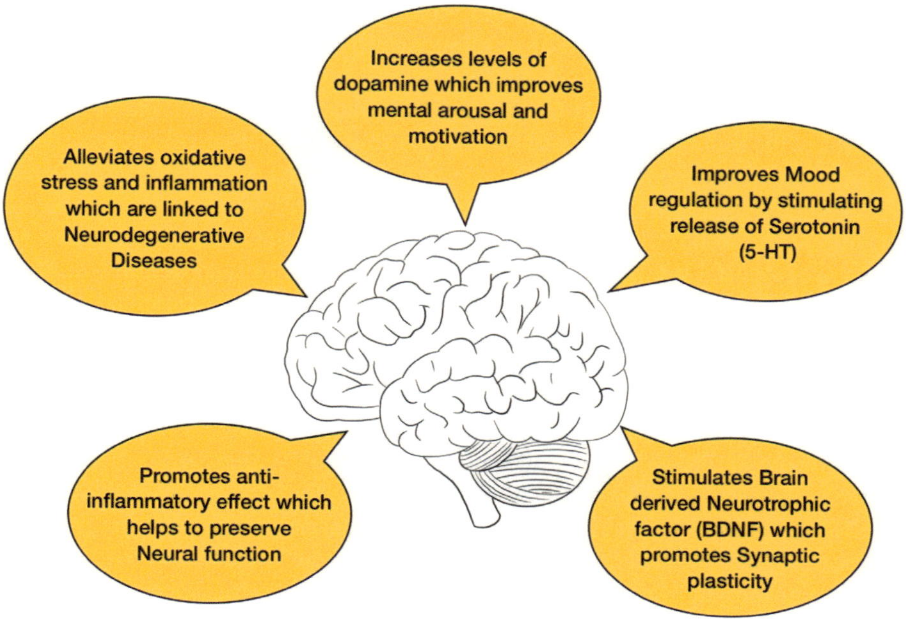

Figure 1.
Influence of aquatic exercises on neurophysiological functions [5].

state. This focused engagement activates brain regions associated with attention and emotional control, such as the prefrontal cortex [6]. Aquatic exercises also facilitate emotional stability by promoting heart rate variability (HRV), an indicator of stress resilience. Controlled breathing in water—especially during activities like water aerobics or swimming—helps regulate the autonomic nervous system, reducing anxiety related distress and enhancing emotional balance. Group aquatic activities create a supportive and inclusive environment that fosters both social and emotional well-being. Participating in classes such as aqua aerobics or synchronised swimming promotes a sense of belonging, which is critical for mental resilience. Shared goals and interactions in a non-competitive, water-based setting encourage cooperation and emotional connection. Studies show that social bonding during group activities can increase oxytocin levels, a hormone associated with trust and emotional warmth [7]. The integration of aquatic exercises into mental resilience programmes leverages water's unique properties to positively impact neurotransmitter activity, focus, emotional stability, and social well-being.

3. Injury prevention: A new dimension in training

Injury prevention is a cornerstone of effective rehabilitation and training, especially for athletes and individuals involved in physically demanding activities. Incorporating hydrotherapy or aquatic exercises into training regimens introduces a scientifically proven method to minimise injury risk while enhancing musculoskeletal strength and joint health. Water's buoyancy significantly reduces the gravitational forces acting on the body, alleviating stress on weight-bearing joints, bones, and soft tissues. This property makes aquatic exercises inherently safer than many land-based activities [2].

For instance, deep water running decreases the impact on knees and hips by up to 90% compared to treadmill running, allowing athletes to maintain cardiovascular fitness without risking overuse injuries [8]. The buoyant environment is particularly beneficial for those recovering from injuries, as it enables controlled, pain-free movement while reducing the risk of re-injury. From a kinesiological perspective, water's viscosity creates multidirectional resistance, which engages stabilising muscles which are often neglected in traditional training (**Figure 2**). This consistent resistance enhances core stability, muscle balance, and neuromuscular coordination, all of which are essential for preventing injuries. Unlike conventional weight training, where resistance is gravity-dependent, water resistance adapts to the speed and force of movement, making it suitable for various fitness levels and therapeutic needs [10]. Bio mechanically, aquatic exercises improve joint alignment and reduce compressive forces. Hydrostatic pressure enhances circulation and reduces inflammation, promoting oxygen and nutrient delivery to tissues, thereby accelerating recovery and supporting cartilage repair. For ageing populations or individuals with degenerative conditions like osteoarthritis, these effects contribute to improved joint function and longevity (**Table 1**). Research in sports medicine highlights that aquatic strength training enhances muscular endurance and power while minimising the risk of overuse injuries [13]. By integrating hydrotherapy into athletic training, practitioners can provide a safe, effective, and adaptable approach to injury prevention, rehabilitation, and overall performance optimisation. Exercises like deep water running, water specific

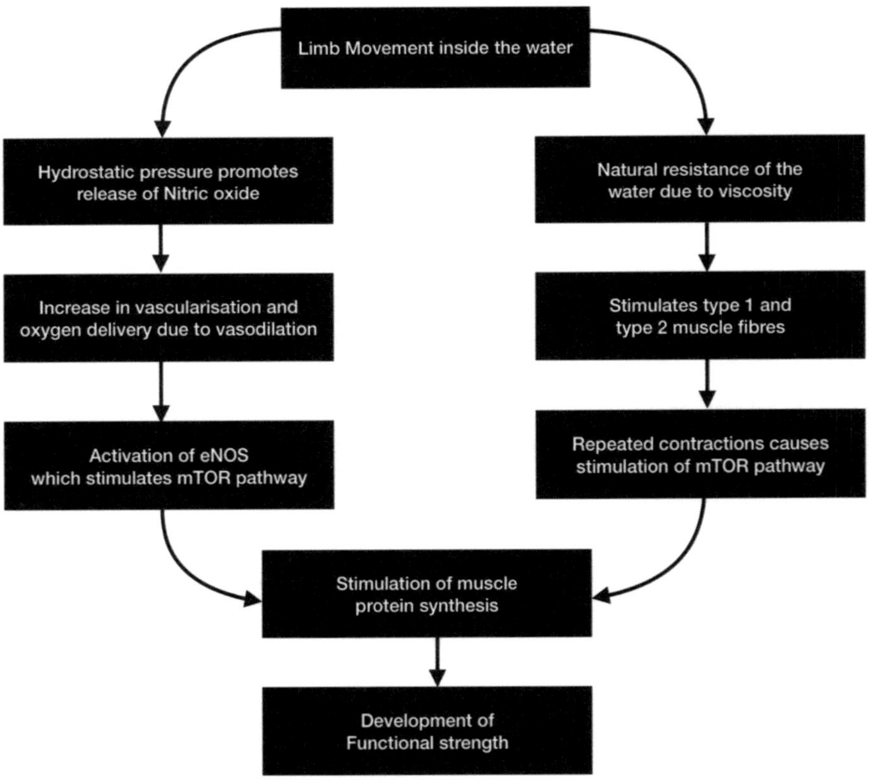

Figure 2.
Physiological mechanisms underlying functional strength improvements through aquatic exercise [9]. mTOR pathway: mammalian target of rapamycin pathway, eNOS: endothelial nitric oxide synthase.

Elements	Mechanism
Reduced Joint loading	Buoyancy minimises pressure over articular surfaces during joint motion.
Increased Joint Range of Motion	Warm water relaxes muscles, hence allows for full range of motion.
Alleviation of pain	Hydrostatic pressure reduces swelling and improves comfort across the articulatory segment.
Enhanced Circulation	Hydrostatic pressure exerted by water enhances vascular perfusion around the joint.
Low-impact environment	The buoyancy of water attenuates mechanical forces during movement thus effectively reducing impact loading on joints.

Table 1.
Key elements contributing to improved Joint Health through Hydrotherapy [11, 12].

therapy, and bad ragaz ring method effectively target major muscle groups without excessive strain on tendons and ligaments, promoting both strength and resilience. Hydrotherapy provides an innovative, low-risk approach to injury prevention by leveraging water's buoyancy, viscosity, and hydrostatic pressure. By reducing impact forces, improving neuromuscular coordination, and promoting joint health, aquatic exercises offer a holistic solution for athletes and active individuals to optimise performance while safeguarding long-term physical health. With its adaptability and proven benefits, hydrotherapy represents a vital tool in modern injury prevention strategies.

Clinical Strategies for Injury Prevention

- Low-impact cardiovascular training: Aquatic exercises like deep water jogging or aqua cycling provide cardiovascular conditioning without the stress of high-impact activities. These activities reduce joint compression, making them suitable for athletes recovering from impact-related injuries or seeking to minimise overuse damage.

- Strength and proprioceptive training: Resistance created by water's viscosity is multidirectional, enhancing muscle engagement, particularly for stabilisers like the rotator cuff and core muscles [13]. This improves joint stability and proprioception, reducing the likelihood of ligamentous injuries, such as sprains or ACL tears.

- Rehabilitation of chronic conditions: Hydrotherapy helps manage overuse injuries such as tendinitis and stress fractures by offering a controlled environment for gradual load progression. For example, exercises targeting the shoulder in water can improve kinetic chain symmetry and mitigate repetitive strain injuries seen in swimmers and throwers.

- Enhancing biomechanical efficiency: Controlled aquatic environments allow athletes to focus on precise movement patterns without compensatory strategies often seen on land. This helps correct neuromuscular imbalances and improve functional alignment, decreasing the risk of injury recurrence [10].

- Hydrotherapy for recovery: The hydrostatic pressure of water reduces swelling, enhances circulation, and aids in neutralising local metabolic acidosis post-exercise, improving recovery times and preparing athletes for subsequent training sessions.

4. Principles of designing aquatic exercise programmes

Aquatic exercise programmes offer a versatile and evidence-based approach to enhancing physical and mental health across diverse populations [14]. Designing effective programmes requires a comprehensive understanding of individual needs, the incorporation of mental health objectives, and a structured system for tracking recovery and progress. This section outlines these principles while drawing on recent scientific research and highly cited literature in sports medicine. Individualised aquatic exercise programmes begin with a detailed assessment of the participant's physical, mental, and functional requirements. Factors such as fitness level, injury status, medical history, and performance goals must be carefully evaluated. Functional movement assessments, flexibility tests, and cardiovascular evaluations are useful tools for creating customised exercise plans [1, 15]. For athletes, specific needs such as sport-specific movement patterns and injury prevention goals should be incorporated. For example, swimmers may focus on shoulder stabilisation, while runners might benefit from lower-body strength and proprioceptive training in water. Tailoring programmes also involves adjusting water depth, temperature, and equipment like resistance paddles or buoyancy belts to match the participant's needs [8, 16]. Aquatic exercise programmes are uniquely suited to addressing mental health objectives alongside physical fitness (**Table 2**). Strategies to enhance mental resilience can include mindful movement sequences, slow-paced stretches, and synchronised breathing exercises in water. Research shows that immersion in warm water promotes relaxation, reduces anxiety, and stimulates the release of mood-enhancing neurotransmitters like serotonin and dopamine [4]. Tailoring sessions to include meditative practices or group activities further supports emotional well-being. For individuals with stress-related conditions, low-impact aquatic yoga or Ai-chi (Aqua form of Tai-chi) movements can enhance the mind-body connection, alleviate stress, and foster mindfulness [1].

Mental health benefits	Effect on physiology	Effect on psychosocial well-being
Decreases psychological stress	Stimulation of Parasympathetic nervous system	Reduces cortisol levels and promotes relaxation
Improves Mood and Emotional well-being	Release of neuro chemicals like endorphins, serotonin etc.	Stimulates the release of endorphins and serotonin, contributing to improved mood and emotional stability
Better sleep quality	Circadian rhythm stabilisation	Improves sleep patterns by reducing anxiety, promoting relaxation, and regulating circadian rhythms
Enhances cognitive function	Improvement in cerebral blood flow and oxygenation of cerebral tissue	Enhances cerebral blood flow and oxygenation, leading to improvements in memory, executive function, and focus
Improves emotional regulation in sensitive cases (PTSD, anxiety)	Alleviates Sensory overload due to buoyancy	Helps individuals with PTSD or anxiety to develop better emotional control and resilience
Alleviates Hyper-arousal state	Alleviates hyper-arousal by Inhibiting Sympathetic nervous system	Reduced hyper-arousal by shifting the balance from a sympathetic (fight-or-flight) to a parasympathetic (rest-and-digest) state

Table 2.
Mental health benefits of aquatic exercise and their impact on psychosocial well-being [17, 18].

Recovery-focused aquatic exercise programmes should account for the participant's stage of healing and rehabilitation. Initial sessions may prioritise low-intensity activities in deeper water to minimise strain on injured areas, progressing to shallower water for more dynamic exercises as strength and mobility improve [19]. Outcome tracking is crucial for evaluating the programme's effectiveness and ensuring continued progress. The following methods can improve the documentation of performance changes, while also contributing to the scientific development of the literature:

1. Functional movement tests: To assess improvements in joint range of motion and strength.

2. Pain scales: To monitor reductions in discomfort during and after sessions.

3. Performance metrics: Such as time to fatigue or strength gains in targeted muscle groups.

Digital health technologies, including wearable devices that track heart rate and movement, can also provide objective data on recovery progress and fitness improvements [8]. Effective Aquatherapy programmes are rooted in a thorough understanding of individual needs, the integration of mental health goals, and the use of systematic tracking for recovery and progress. By leveraging water's unique therapeutic properties, these programmes can address a wide range of physical and psychological objectives, providing participants with a holistic pathway to health and performance.

5. Practical application of aquatic exercises for mental health and fitness in special population

Aquatic exercises are increasingly recognised for their therapeutic benefits, particularly for special populations, including individuals with physical disabilities, chronic conditions, and mental health challenges. These groups often face unique physical and psychological barriers that can hinder their participation in traditional exercise programmes. Aquatic exercise offers an accessible, low-impact alternative, utilising the unique properties of water to improve both mental health and physical fitness. This section explores the practical application of aquatic exercises for mental health and fitness in special populations.

5.1 Aquatic exercises for individuals with disabilities

For individuals with physical disabilities, aquatic exercise offers a safe and effective way to engage in physical activity without the risk of exacerbating pain or injury. Water buoyancy reduces the effects of gravity, allowing for easier movement and less strain on joints and muscles, which is particularly beneficial for individuals with limited mobility or those in wheelchairs [8]. Studies have shown that aquatic exercise can improve strength, flexibility, balance, and cardiovascular health in people with disabilities, contributing to overall quality of life [20]. For example, hydrotherapy has been found to be effective in reducing spasticity and improving range of motion in individuals with cerebral palsy. Water-based rehabilitation programmes also help to improve functional mobility in individuals with spinal cord injuries, providing them with the opportunity to gain greater independence in their daily activities.

5.2 Aquatic exercise for mental health disorders

Aquatic exercises are also effective in improving mental health, as the soothing effects of water contribute to reduction in anxiety, depression, and stress. Immersion in water promotes relaxation by activating the parasympathetic nervous system, lowering cortisol levels, and stimulating the release of neurotransmitters such as serotonin and dopamine [4]. Aquatic exercise programmes that incorporate mindfulness, breathing exercises, and meditative movements have been shown to reduce symptoms of anxiety and depression in individuals with mental health disorders [21]. Water-based interventions have also been associated with improved mood, self-esteem, and social connectedness, especially when combined with group activities. For individuals with conditions such as post-traumatic stress disorder (PTSD) or major depressive disorder, the calming effects of water combined with exercise can provide an effective strategy for improving mental well-being [1].

5.3 Aquatic exercise for older adults

Older adults often experience a decline in muscle strength, joint mobility, and cognitive function, which can be mitigated through aquatherapy. Research has shown that water-based programmes improve balance, flexibility, and muscular endurance in older populations, reducing the risk of falls and injuries. These improvements are particularly beneficial for those with age-related conditions such as osteoarthritis, where water buoyancy helps decrease the load on joints, allowing for pain-free movement. Moreover, aquatic exercise provides an opportunity for older adults to engage in social activities, which has been linked to better mental health outcomes. A study published in the *Journal of Ageing and Physical Activity* found that older adults who participated in water aerobics exhibited greater improvements in cognitive function and emotional well-being compared to those who did not engage in such activities [4].

5.4 Aquatic exercises for individuals with chronic illnesses

Chronic conditions such as arthritis, fibromyalgia, and cardiovascular disease can significantly impact an individual's ability to engage in traditional exercise. Aquatic exercises provide a low-impact alternative that reduces the strain on affected joints while promoting cardiovascular health, muscle strength, and flexibility. For instance, hydrotherapy has been shown to reduce pain and improve mobility in individuals with rheumatoid arthritis [2]. A review published in the Journal of Sports Sciences concluded that aquatic exercise can improve functional capacity, reduce pain, and enhance the quality of life for individuals with chronic conditions [15]. Additionally, water-based therapy has been found to improve the overall physical and psychological well-being of individuals with fibromyalgia, reducing fatigue and enhancing mood [20].

5.5 Practical considerations for implementing aquatic exercise Programmes for special populations

When designing aquatic exercise programmes for special populations, it is essential to tailor the activities to the specific needs and abilities of participants. Factors such as water depth, temperature, and exercise intensity should be considered. Shallow water exercises may be suitable for improving balance and strength, while deep water exercises can be used to reduce joint stress and focus on cardiovascular endurance. Additionally, programmes should incorporate flexibility, strength training, and aerobic exercises to provide a well-rounded fitness regimen. For those with mental health concerns, mindfulness and breathing techniques should be integrated to promote emotional well-being.

Safety precautions, such as ensuring participants can swim or providing flotation devices, should also be in place, especially for individuals with limited mobility or fear of water. Aquatic exercise represents an effective and adaptable form of physical activity that addresses the unique needs of special populations. Whether for individuals with physical disabilities, mental health challenges, chronic illnesses, or ageing-related conditions, aquatic exercises offer significant benefits for improving both physical fitness and mental well-being.

6. Challenges and myths about aquatic therapy

Aquatic therapy has proven benefits for rehabilitation, fitness, and mental well-being. However, its widespread adoption is often hindered by misconceptions, accessibility issues, and environmental constraints. This section explores common myths surrounding aquatic therapy, identifies barriers to its implementation, and provides strategies to overcome these challenges, ensuring broader accessibility.

6.1 Common misconceptions about aquatic exercises

1. *Myth*: Aquatic exercises are only for older adults or individuals with disabilities.

   While aquatic therapy is frequently recommended for rehabilitation and ageing populations, it is equally effective for athletes, children, and individuals seeking general fitness improvement. Research demonstrates that water-based exercises can enhance strength, endurance, and flexibility across all age groups and fitness levels [1, 2].

2. *Myth*: Aquatic exercises are not intense enough for fitness or weight loss.

   Many assume that aquatic workouts are low-intensity; however, water resistance provides significant muscular engagement. High-intensity aquatic exercises, such as aquatic plyometrics or water aerobics, are comparable to land-based training in calorie expenditure and cardiovascular benefits [13].

3. *Myth*: Participants must know how to swim to benefit from aquatic therapy.

   Aquatic therapy often takes place in shallow pools or uses flotation devices, making it accessible even to non-swimmers. Proper guidance from trained professionals ensures safety and effectiveness [22].

6.2 Barriers to accessibility and how to overcome them

1. Physical accessibility:

   Individuals with disabilities or limited mobility may find it challenging to access aquatic facilities due to inadequate infrastructure. Ensuring facilities are equipped with ramps, lifts, and adaptive equipment can significantly improve inclusivity. The Americans with Disabilities Act (ADA) provides guidelines for pool accessibility, which can serve as a global standard for improvement [1].

2. Financial barriers:

   The cost of constructing and maintaining aquatic facilities can be prohibitive, limiting access in lower-income areas. Addressing these challenges may involve collaboration with local governments or NGOs to subsidise aquatic programmes, particularly for underserved populations [13].

3. Cultural and psychological barriers:

   Cultural taboos, fear of water, or lack of awareness may deter participation. Community outreach programmes, educational initiatives, and introductory sessions can help overcome these barriers. For example, water confidence classes can address Aqua phobia, promoting gradual acclimatisation [23].

6.3 Adapting water therapy for varied environmental and facility constraints

1. Limited pool space:

   Small or crowded facilities can still offer effective aquatic therapy by using creative programme designs such as circuit training or limiting group sizes. Portable resistance equipment can further enhance training in compact spaces.

2. Seasonal or climate restrictions:

   Outdoor aquatic programmes face challenges during colder months. Utilising indoor heated pools or implementing mobile pool solutions can provide year-round access. Innovations like thermal pool covers and portable pool heaters also extend the usability of outdoor facilities.

3. Water resource management:

   In areas with water scarcity, sustainable practices such as recycling pool water or utilising natural water bodies with appropriate safety measures can help maintain aquatic programmes.
   As a conclusion, Aquatic therapy faces challenges ranging from misconceptions to logistical barriers, but with proper education, adaptive strategies, and investment, these obstacles can be mitigated. Addressing these issues not only broadens accessibility but also highlights the versatility and inclusivity of aquatic therapy.

7. Future of aquatic therapy

The future of aquatic therapy is being shaped by technological advancements, innovative programming, and a focus on inclusivity. These developments are making aquatic exercises more effective, accessible, and appealing, especially to underserved populations. Ongoing research focuses on developing standardised protocols for specific injuries or conditions, ensuring consistency and effectiveness in therapeutic outcomes. The integration of these advancements and strategies positions aquatic therapy as a vital component of future rehabilitation and wellness programmes.

7.1 Technological innovations: Virtual reality and wearable devices in the pool

1. Virtual reality (VR) in aquatic therapy:

   Virtual reality is transforming rehabilitation and fitness in aquatic environments by offering engaging, gamified experiences. VR systems tailored for water activities, such as waterproof headsets, enable users to simulate various scenarios like walking on different terrains or performing sport-specific drills. This approach improves motivation, adherence, and cognitive engagement during therapy [24]. Studies have also shown that VR-based aquatic exercises can enhance motor control, reduce pain perception, and improve psychological well-being in individuals recovering from injuries or neurological conditions [25].

2. Wearable devices for monitoring performance:

   The use of waterproof wearable technology, such as fitness trackers and heart rate monitors, allows real-time monitoring of physiological parameters like heart rate, calorie expenditure, and muscle activity during aquatic sessions. These devices enable therapists to customise sessions based on the client's response, track progress over time, and ensure safety during high-intensity workouts [26]. Advanced sensors that monitor biomechanical movements in water are also being integrated into rehabilitation programmes to refine movement patterns and prevent injuries.

3. Smart pool systems:

   Smart pools equipped with automated water resistance controls, underwater cameras, and digital feedback screens are emerging as valuable tools for aquatic therapy. These systems can analyse biomechanics, provide visual feedback, and customise resistance levels to match individual needs, enhancing the precision and effectiveness of therapy sessions [27].

7.2 Promoting aqua exercises in underserved communities

1. Addressing financial barriers:

   Communities with limited access to aquatic facilities face significant barriers due to high costs associated with pool construction and maintenance. Partnerships between governments, NGOs, and private organisations can fund mobile pools, low-cost aquatic programmes, or public access initiatives to bring aquatic therapy to underserved areas. Initiatives like the Make a Splash Foundation have successfully expanded swimming and aquatic therapy programmes in low-income communities [20].

2. Cultural inclusivity:

   Promoting awareness about the benefits of aquatic therapy through culturally tailored outreach programmes can increase acceptance. Educating underserved populations about the mental and physical health benefits of aquatic therapy, including its role in reducing chronic conditions like arthritis and hypertension, is critical. Training local therapists to provide cost-effective programmes in community centre's or natural water bodies can also extend access.

3. Environmental sustainability:

   Using eco-friendly solutions, such as solar-powered heating systems or natural pools, can make aquatic therapy more sustainable and accessible in resource-scarce regions. Additionally, portable aquatic therapy equipment, such as inflatable pools or modular swim systems, offers cost-effective options for delivering water-based programmes.

4. Telehealth and virtual training programmes:

   The integration of telehealth with aquatic therapy allows therapists to remotely guide individuals and communities in establishing safe and effective aquatic exercise routines. Virtual platforms can provide instructional content, monitor progress, and offer support to underserved areas lacking access to therapists or facilities [21]. The future of aquatic therapy lies in harnessing technology to enhance effectiveness and expand accessibility to benefit diverse populations. Innovations like VR, wearable devices, and smart pools, combined with inclusive strategies for underserved communities, are paving the way for a more equitable and impactful approach to water-based rehabilitation and fitness.

8. Empowering life through water

Aquatic therapy stands as a profound testament to the healing and transformative power of water. As highlighted throughout this chapter, water-based exercises offer a unique synergy of physical rehabilitation, mental well-being, and fitness improvement, suitable for diverse populations. From addressing injury prevention and promoting mental resilience to fostering inclusivity in underserved communities, aquatic therapy has emerged as a cornerstone of holistic health and wellness.

1. Prioritise water-based therapies in health strategies

   Policymakers, healthcare providers, and fitness professionals must integrate aquatic therapy into mainstream health initiatives. Its proven benefits for cardiovascular health, musculoskeletal recovery, and mental well-being can significantly complement traditional therapeutic methods.

2. Promote accessibility for All

   Bridging gaps in access to aquatic facilities is essential to ensure that water therapy benefits reach individuals from all socio-economic backgrounds. Investments in mobile pools, partnerships with NGOs, and subsidised programmes can make aquatic therapy universally available.

3. Harness technology to enhance reach and impact

   Innovations such as virtual reality, wearable devices, and telehealth platforms can revolutionise how aquatic therapy is delivered. These advancements not only enhance programme efficacy but also expand access to individuals in remote or underserved regions.

4. Embrace water as a tool for mental resilience

   The therapeutic properties of water provide a unique environment for reducing stress, improving focus, and fostering emotional well-being. Regular participation in aquatic exercises can empower individuals to build mental resilience and lead healthier lives, making it an essential component of modern mental health strategies.

9. Conclusion

Water's unique therapeutic properties make aquatic therapy a vital yet underutilised tool in modern rehabilitation. Its buoyancy reduces joint load, enabling early movement in injury recovery, while water's resistance enhances muscle strength, balance, and proprioception. These qualities allow physiotherapists to address acute and chronic conditions, from post-surgical rehabilitation to managing osteoarthritis and neurological disorders. Aquatic therapy bridges the gap between traditional rehabilitation and holistic health by offering a low-impact, adaptable modality for patients across various conditions and fitness levels. Physiotherapists are encouraged to integrate hydrotherapy into their practice by pursuing specialised training. This not only broadens their professional expertise but also provides patients with an evidence-based approach to recovery and wellness. As the field evolves, leveraging advanced technologies such as wearable sensors and adaptive hydrotherapy pools can enhance treatment outcomes. Promoting access through community programmes and education is equally essential to breaking barriers of affordability and misconception. By embracing aquatic therapy, physiotherapists can unlock a powerful patient centric tool that complements traditional care, inspiring a new standard in holistic rehabilitation and empowering patients to achieve lasting physical and mental well-being.

Acknowledgements

I am deeply grateful to Aquacentric, Worli, for providing invaluable exposure to aquatic therapy and to my patients, whose journeys have been my greatest source of learning. Heartfelt thanks to my parents and friends for their unwavering support, and A special mention goes to Komal A. Sangani for being a true companion in my journey toward becoming a specialist in sports physiotherapy. Your encouragement and camaraderie have been pivotal in my personal and professional growth. This chapter is a reflection of all the encouragement and inspiration I have received along the way to make a meaningful impact. The author acknowledges the use of ChatGPT 4.1 for the language polishing of the manuscript.

Author details

Aryan Shukla
Senior Sports Physiotherapist, Aquacentric, Mumbai, India

*Address all correspondence to: dr.muscle960@icloud.com

References

[1] Becker BE. Aquatic therapy: Scientific foundations and clinical rehabilitation applications. PM&R. 2009;**1**(9):859-872

[2] Hinman RS, Heywood SE, Day AR. Aquatic physical therapy for hip and knee osteoarthritis: Results of a single-blind randomized controlled trial. Physical Therapy. 2007;**87**(1):32-43

[3] Wilcock IM, Cronin JB, Hing WA. Physiological response to water immersion: A method for sport recovery? Sports Medicine. 2006;**36**(9):747-765

[4] Cochrane DJ, Vella CA, Cain DP. The effects of water immersion on mood states in athletes. Journal of Sport Rehabilitation. 2013;**22**(3):225-234

[5] Oh S, Lee S. Effect of aquatic exercise on physical function and QOL in individuals with neurological disorder: A systematic review and meta-analysis. Journal of Bodywork and Movement Therapies. 2021;**27**:67-76. DOI: 10.1016/j.jbmt.2021.01.009. Epub 2021 Jan 21

[6] Uvnas-Moberg K, Arn I, Theorell T. The psychobiology of oxytocin: How it helps us create and maintain relationships. Frontiers in Psychology. 2014;**5**:1049

[7] Nagle JP et al. The neural correlates of mindfulness practice. Journal of Neuroscience Research. 2015;**93**(12):172-178

[8] Silvers HJ, Dolny DG. Comparison of land-based and aquatic-based plyometric programs for improving power and vertical jump in adolescent athletes. Journal of Strength and Conditioning Research. 2011;**25**(8):2250-2256

[9] Kobayashi J, Uchida H, Kofuji A, Ito J, Shimizu M, Kim H, et al. Molecular regulation of skeletal muscle mass and the contribution of nitric oxide: A review. FASEB Bioadvances. 2019;**1**(6):364-374. DOI: 10.1096/fba.2018-00080

[10] Colado JC, Tella V, Triplett NT. A method for monitoring intensity during aquatic resistance exercises. Journal of Strength and Conditioning Research. 2010;**24**(1):204-210

[11] Lei C, Chen H, Zheng S, Pan Q, Xu J, Li Y, et al. The efficacy and safety of hydrotherapy in patients with knee osteoarthritis: A meta-analysis of randomized controlled trials. International Journal of Surgery. 2024;**110**(3):1711-1722. DOI: 10.1097/JS9.0000000000000962

[12] Bartels EM, Lund H, Hagen KB, Dagfinrud H, Christensen R, Danneskiold-Samsøe B. Aquatic exercise for the treatment of knee and hip osteoarthritis. Cochrane Database of Systematic Reviews. 2007;**4**:CD005523. DOI: 10.1002/14651858. CD005523.pub2. Update in: Cochrane Database Syst Rev. 2016 Mar 23;3:CD005523. DOI: 10.1002/14651858.CD005523.pub3

[13] Reilly T, Dowzer CN, Cable NT. The physiology of deep-water running. Journal of Sports Sciences. 2003;**21**(12):959-972

[14] Prins J, Murata NM. The benefits of aquatic therapy for rehabilitation and training. International Journal of Aquatic Research and Education. 2008;**2**(2):162-170

[15] Robinson LE et al. Effects of aquatic exercise on quality of life in individuals with arthritis. American Journal of Health Promotion. 2010;**24**(4):274-278

[16] Bressel E, Dolny DG. Aquatic plyometric training increases vertical jump in female volleyball players. Medicine and Science in Sports and Exercise. 2011;**43**(10):1811-1819

[17] Silva LAD, Tortelli L, Motta J, Menguer L, Mariano S, Tasca G, et al. Effects of aquatic exercise on mental health, functional autonomy and oxidative stress in depressed elderly individuals: A randomized clinical trial. Clinics (São Paulo, Brazil). 2019;**74**:e322. DOI: 10.6061/clinics/2019/e322. Epub 2019 Jun 27

[18] Tang Z, Wang Y, Liu J, Liu Y. Effects of aquatic exercise on mood and anxiety symptoms: A systematic review and meta-analysis. Frontiers in Psychiatry. 2022;**13**:1051551. DOI: 10.3389/fpsyt.2022.1051551

[19] Vaile J, Gill N, Blazevich A. The effect of contrast water therapy on symptoms of delayed onset muscle soreness. Journal of Strength and Conditioning Research. 2007;**21**(3):697-702

[20] USA Swimming Foundation. Expanding aquatic access: The role of community partnerships. Journal of Community Health. 2021;**46**(2):329-335

[21] Waller S, Blustein J, Spangler J. Telehealth delivery of aquatic therapy: Overcoming barriers in underserved populations. International Journal of Aquatic Research and Education. 2020;**12**(1):22-30

[22] Kemp J, Williams S, Smith L. Aquatic exercise for mental health: A review of the literature. Journal of Psychiatric Research. 2018;**103**:64-72

[23] Waller S. Aquatic exercise and rehabilitation for individuals with disabilities: Benefits and considerations. Journal of Rehabilitation Research and Development. 2010;**47**(2):71-80

[24] Hoffmann M, Steinhagen-Thiessen E, Silberhorn T. Virtual reality in aquatic therapy: Applications and benefits. Journal of Neuroengineering and Rehabilitation. 2021;**18**(4):44-53

[25] Laver K, George S, Thomas S, Deutsch JE, Crotty M. Virtual reality for stroke rehabilitation. Cochrane Database of Systematic Reviews. 2020;**11**(11):CD008349

[26] McManus AM, Burns SF, Ghio K. Advances in wearable technology for aquatic fitness and rehabilitation. Sports Medicine. 2022;**52**(5):777-789

[27] Finkelstein S, Lapshin O, Castro E. Innovations in aquatic therapy: Utilizing smart pool systems for enhanced rehabilitation. Journal of Rehabilitation Research and Development. 2019;**56**(3):101-115

Chapter 10

Perspective Chapter: Physical Activity as Treatment for Chronic Diseases – Guideline for Prescription

Eugénie d'Alessandro

Abstract

Physical activity is an essential component of the therapeutic management of chronic diseases. It contributes significantly to the prevention of complications and recurrence of these diseases. The prescription of physical activity must therefore find its place in the daily practice of doctors. However, this prescription remains too marginal. In this review, we propose a systematic approach to integrate the prescription of physical activity for therapeutic purposes into the daily practice of doctors. Based on international recommendations and standardized tools, we have developed a step-by-step procedure that allows to identify the target patients, carry out the medical check-up and write the physical activity prescription. The aim of this guideline is to integrate physical activity as treatment of chronic diseases more systematically and as early as possible into patient's care.

Keywords: physical activity, chronic diseases, medical prescription, FITT model, non-drug therapy

1. Introduction

The health benefits of physical activity are well established. Many research, reports and recommendations have been published on the health benefits of physical activity. Practiced on a regular basis, physical activity improves the physical condition, autonomy, health and quality of life of populations at all ages [1, 2]. Physical activity is a primary prevention tool by limiting the risk of chronic diseases. But it is also considered a non-drug therapy in its own right in many chronic diseases. It can be prescribed alone as a first-line treatment for type 2 diabetes, uncomplicated hypertension, non-severe COPD, overweight and obesity, a mild-to-moderate depressive episode, as well as in elderly people at risk of falls or loss of autonomy. It can also be prescribed in combination with drug therapies in cancer, cardiovascular diseases or chronic inflammatory rheumatic diseases [1–4].

However, in our societies, the levels of sedentary lifestyle and inactivity continue to increase [5–7]. Sedentary time significantly increase cardiovascular, cancer and

> Physical activity is defined as any body movement produced by skeletal muscle, which results in a higher energy expenditure than that of resting metabolism. Physical activity includes activities carried out in the home, work, travel and leisure contexts (including sport).
>
> The medical prescription of adapted physical activity should allow patients to adopt a physically active lifestyle on a regular basis.
>
> In sport, practitioners follow rules. The goal is performance. In adapted physical activity, it is the patient's rules that guide the practice. It is no longer a question of performance or progression. Physical activity must therefore be adapted to the taste, physical condition and state of health of each patient.

Box 1.
Definition of physical activity.

mortality risk. Sedentary behavior in combination with chronic diseases increases all-cause mortality risk [8–10]. Physical inactivity is the fourth leading risk factor for mortality worldwide (6% of deaths), just after high blood pressure, smoking and blood glucose levels [11]. According to the WHO, in 2022, nearly one-third (31%) of adults worldwide, or about 1.8 billion people, were inactive, meaning they did not meet physical activity recommendations. This level of physical inactivity increased by 5% between 2010 and 2022. And as WHO has strengthened, if the trend continues, the level of inactivity is expected to increase further to 35% by 2030. The highest rates of physical inactivity were observed in the Asia-Pacific region with high income (48%) and South Asia (45%). Inactivity levels in other regions range from 28% in high-income Western countries to 14% in Oceania [7].

Faced with these risk factors, physical activity (see **Box 1**) is an essential component of primary prevention to fight against physical inactivity and sedentary lifestyle and to prevent the occurrence of pathologies [3, 12]. But physical activity is also of major interest in therapy for the management of chronic diseases. Chronic diseases promote a psychophysiological process of physical inactivity with physical deconditioning, sedentary lifestyle, loss of confidence in these physical abilities and minor depressive disorders. Practiced in appropriate conditions, regular physical activity contributes to the treatment of these diseases and reduces complications and recurrences that are essential issues for maintaining quality of life and autonomy [13, 14]. Physical activity reduces overall mortality by 50% in breast, prostate and colon cancers regardless of the stage of progression at which the cancer was diagnosed. In addition, after a first cardiovascular accident, physical activity leads to a 30% reduction in cardiovascular mortality, a 26% decrease in overall mortality and a 31% reduction in rehospitalization [3]. Another example, physical activity is the only treatment that has shown effectiveness in relieving fatigue during chemotherapy. In short, regular physical activity is one of the main therapeutic responses for the management of chronic diseases.

2. What is the use for the prescription of physical activity?

The benefits of practicing physical activity, adapted if necessary, undoubtedly outweigh the risks involved, regardless of the age and state of health of the person. The role of physical activity in the therapeutic management of chronic diseases is therefore essential. However, it is not a question of medicalizing the practice of physical activity. This requirement concerns a targeted audience.

First, this chapter does not deal with rehabilitation, which must be the subject of a specific prescription and care by physiotherapists. Nor do we address the issue of sports activities practiced at high intensities in leisure or competition. Even people with chronic pathologies can exercise at a high intensity. A person with well-balanced diabetes and a good physical condition can compete very well. In this case, the medical supervision consists above all in eliminating any contraindications. The prescription of physical activity itself has, in this case, a limited place since the person concerned and his or her sports educator will decide on the physical program to be implemented.

Recommendations concerning the prescription of physical activity for health purposes are presented here for adults, the elderly and people with disabilities, with chronic diseases or with risk factors (sedentary lifestyle, loss of autonomy, physical deconditioning, etc.) and with a fragile physical condition. This prescription is therefore intended for people for whose high intensities are not sought. The objective is the regularity of physical activity, which is the only criterion to be taken into account to obtain an effect on health. This physical activity must be adapted to the tastes and needs of the patients, and respect the basic rules: hydration, warm-up before the session and recovery at the end of the session.

It is therefore not a question of medicalizing physical activity. Many people, including people with chronic diseases, engage in regular physical activity without specific medical supervision. However, in sedentary and insufficiently active people, several studies tend to show that a medical prescription and the support it requires are factors of motivation and adherence. The prescription of physical activity therefore has its place in the daily practice of doctors.

In this review, we propose a systematic approach to integrate the prescription of physical activity for therapeutic purposes into the daily practice of doctors. Based on international recommendations and standardized tools [1, 12, 15–17], we have developed a step-by-step procedure that allows to identify the target patients, carry out the medical check-up and write the physical activity prescription.

3. Prescribing physical activity: A systematic approach

As with the prescription of drug treatments, the prescription of physical activity must specify indications, "dosages," precautions and possible contraindications according to the patients' profiles. We propose here a systematic approach to easily integrate the prescription of physical activity into daily medical practice.

3.1 Step 1: Identifying patients at risk

All patients who request it or for whom the doctor deems it useful can benefit from a medical consultation to start or resume physical activity.

However, in some situations, medical advice is more urgently needed. We propose to use a standardized screening questionnaire: The Physical Activity Readiness Questionnaire for Everyone [18]. The patient can answer this questionnaire alone or with the help of his doctor. This questionnaire makes it possible to identify patients for whom a medical check-up and a prescription for physical activity are necessary.

3.2 Step 2: To look for severe acute pathologies that have not been stabilized

In this second step, the objective is to look for a possible pathology with an unknown and/or unstable cardiovascular risk in the patient (see **Box 2**). These pathologies do not constitute an absolute contraindication to physical activity.

These pathologies require a specialist advice before prescribing physical activity, even at moderate intensity:

- Instable angina pectoris
- Decompensated heart failure
- Complex ventricular rhythmic disorders
- Severe uncontrolled high blood pressure
- Pulmonary arterial hypertension (> 60 mmHg)
- Presence of large or pedicled intracavitary thrombus
- Myocarditis or acute pericardial effusion
- Severe obstructive cardiomyopathy
- Tight and/or symptomatic aortic stenosis
- Recent thrombophlebitis with or without pulmonary embolism
- Diabetes with foot ulcer for physical activities that solicit the limbs inférieurs
- Uncontrolled acute respiratory conditions
- Uncontrolled diabetes with ketonuria/ketosis
- Significant undernutrition
- Extreme fatigue or obvious physical disability
- Marked anemia (hemoglobin <8 g/L)
- Pressure ulcer and progressive wound depending on severity and location

Box 2.
Specialist medical advice in the event of a severe acute non-stabilized pathology.

On the contrary, physical activity contributes to the treatment of these diseases and the prevention of their complications. However, before starting physical activity, these pathologies must be identified and stabilized.

3.3 Step 3: Assess cardiovascular risk and the indication for a stress test

The cardiovascular risk level is the probability of a major cardiovascular event (heart attack, sudden death) occurring in a person over a given period. The cardiovascular risk classification allows patients to be divided into four risk levels: low, moderate, high and very high. To assess the cardiovascular risk in his patient, the doctor must consider:

- The SCORE charts [19],
- Non-SCORE risk factors that may influence the SCORE,
- Chronic pathologies known to be at high or very high cardiovascular risk,
- An age of more than 65.

Depending on the patient's level of cardiovascular risk and the level of intensity envisaged for the practice of physical activity, a stress test may be indicated, as well as a cardiology advice.

See **Tables 1** and **2**.

Very high	SCORE ≥ 10% Myocardial infarction and coronary artery disease Peripheral vascular diseases (PAD, aortic aneurysm, presence of significant atherosclerotic plaques) Cerebrovascular accident and transient ischemic attack Severe chronic kidney disease Diabetes with macrovascular or microvascular complications (including proteinuria) or associated risk factor (smoking, hypertension, hypercholesterolemia)
High	5% < SCORE < 10% Moderate chronic kidney disease HTA > 180/110 mmHg Total hypercholesterolemia>3.1 g/L Other diabetes Cancer and history of cancer and chemotherapy with cardiovascular toxicity Autoimmune and inflammatory diseases Sleep apnea Preeclampsia, pregnancy-related hypertension, polycystic ovarian syndrome Obesity: BMI > 30 kg/m2 or abdominal circumference > 102 cm for men or 88 cm for women Age > 65 years
Moderate	1% < SCORE < 5%
Low	SCORE < 1%

Table 1.
Level of cardiovascular risk.

Starting or continuing physical activity with a level of intensity	Low cardiovascular risk	Moderate cardiovascular risk	High and very high cardiovascular risk
Low intensity	ST not recommended	ST not recommended	ST not recommended
Moderate intensity	ST not recommended	ST not recommended	ST not recommended
High intensity	ST not recommended	Possible ST	Recommended ST

Stress test not recommended: The doctor should not prescribe the stress test unless there is a special case duly justified.
Possible stress test: The stress test can be carried out if it is justified by a particular context, but the examination should not be prescribed systematically.
Recommended stress test: The doctor must have the stress test performed, except in special clinical situations.

Table 2.
Indication for Stress Test (ST) by Cardiovascular Risk.

3.4 Step 4: Assess the need for additional assessments

Depending on the patient's pathologies, certain additional examinations and specialist advice may be considered. They must not be systematic. As we have already explained, when practiced at moderate intensities, physical activity almost always has a favorable benefit-risk balance. The indications for these additional examinations must therefore remain targeted according to the clinical context. They are largely based on the clinical judgment of the physician.

Among the additional check-ups, the following may be considered:

- A pulmonary advice: with pulmonary function tests or a six-minute walk test to look for desaturation on exertion, in particular.
- An ophthalmological advice: for the search for retinopathy or glaucoma.
- A biological assessment: with blood count, blood glucose, lipid profile, serum creatinine.

3.5 Step 5: Prescribing physical activity, general indications

At the end of the medical check-up, the physical activity prescription must specify the general indications. As in drug treatments for which a dosage is predefined, these indications follow international recommendations.

According to the FITT method, they are divided into four components, namely, Frequency, Intensity, Type and Time.

3.5.1 Frequency

The recommendations are 150 to 300 minutes of moderate-intensity physical activity per week. These 150 to 300 minutes should be spread over the week, with sessions of 20 to 60 minutes 3 to 7 times a week.

3.5.2 Intensity

Aerobic endurance physical activity (moderate intensity) is the general recommendation. For the patient, this corresponds to physical activity that causes moderate shortness of breath. On the Borg Rating of Perceived Exertion (RPE) scale scored from 1 to 10, the patient should remain between 1 and 5.

Special cases:

- Patients who must respect the maximum symptomatic time aim for intensities of 1 to 3-4 on the Borg scale.
- Patients treated for cancer should reach moderate intensities of 4 to 6-7 on the Borg scale (excluding cardiovascular contraindications).

3.5.3 The type

For all patients, it is necessary to prescribe endurance, strengthening, flexibility, balance and coordination exercises.

Depending on the state of health, certain components will be more important, such as strengthening in sarcopenia, cardiovascular pathologies or chronic obstructive pulmonary disease, balance exercises for older adult fall prevention, coordination in neurodegenerative pathologies.

3.5.4 Time

Physical activity must be practiced throughout life. As part of the prescription for entry into an adapted physical activity program, the prescription can cover 1 year. It will be renewed annually.

In these general indications, two points deserve to be emphasized: respect for a moderate intensity (aerobic endurance) and respect for the rule of no pain.

3.5.5 Moderate-intensity physical activity

Apart from the exceptional contraindications listed in **Box 2**, the benefit-risk balance of physical activity is always favorable as long as this physical activity is done at moderate intensity. This message must be very clear to the patient. Physical activity is still too often perceived as requiring intense effort, surpassing oneself. Patients must therefore understand that they do not need to go to high levels of effort for physical activity to have an effect on their health. The essential condition is the regularity of the practice and not its intensity.

More than heart rate, which is not an obvious measure for all patients, the Borg Rating of Perceived Exertion scale is the simplest tool to assess and monitor this level of effort by the patient himself. It is therefore necessary to attach to the prescription a document indicating the level of effort to be respected. For most patients, this level is between 1 and 5 on the Borg RPE scale.

A prescription that includes the Borg Rating of Perceived Exertion scale [20] and good practice recommendations (like the 10 golden rules of the French Federation of Cardiology [21]) helps ensure a good understanding of the patient (see **Box 3**). It is a bit like the "Patient Medication Information" proposed by the FDA for prescription drug products. It provides clear, concise, accessible and useful information and it participates in the therapeutic education of the patient. Therefore, it contributes to the safety of the practice.

3.5.6 The rule of no pain

Strengthening exercises that are too intense and too repeated could aggravate pathologies of the musculoskeletal system such as tendinopathy, osteoarthritis, inflammatory disorders or prostheses. However, these pathologies do not

1. I report to my doctor any chest pain or abnormal shortness of breath that occurs with exertion.
2. I report to my doctor any heart palpitations that occur during exercise or just after exercise.
3. I report to my doctor any discomfort occurring during exercise or just after exercise.
4. I always respect a 10-minute warm-up and cool-down during my sports activities.
5. I drink 3 to 4 sips of water every 30 minutes of exercise in training and competition.
6. I avoid intense activities in outdoor temperatures below −5°C or above 30°C and during pollution peaks.
7. I do not smoke, at least never in the 2 hours before and after my sports activity.
8. I never use doping substances, and I avoid self-medication in general.
9. I do not do intense sports if I have a fever, or in the 8 days following a flu episode (fever—body aches).
10. I have a medical check-up before resuming intense sports activity if I am over 35 years old for men and 45 years old for women.

Box 3.
The 10 golden rules for safe physical activity.

contraindicate mobilization. On the contrary, to remain functional, a joint must retain its amplitude and flexibility, its musculature and proprioception. It is therefore necessary to mobilize the joints that are suffering.

To avoid injuries and aggravations, the rule is simple: The exercises offered must not trigger or exacerbate pain. This criterion must therefore be mentioned in the prescription.

Depending on the severity of the damage, therapeutic education of the patient and a prescription for physiotherapy for more specific work (analgesic massage, physiotherapy, passive amplitude work, etc.) may be combined.

3.6 Step 6: Prescribing physical activity, specific indications

3.6.1 Specific indications according to drug treatments

Certain drug treatments can increase certain risks during physical activity such as dehydration, hypoglycemia, hypotension or thermoregulation disorders or electrolyte disorders. In a way, these are drug interactions to be taken into account to limit the risk of adverse events.

See **Table 3.**

Drugs	Potential adverse effects	Precautions to be mentioned in the prescription of physical activity
Beta blockers	Alter the ability to assess exercise intensity by heart rate. May promote the occurrence of hypoglycemia, especially if combined with insulin or insulin secretors. Possible thermoregulation disorders. Risk of post-exercise hypotension.	Evaluation of exercise intensity by perceived sensations, especially respiratory sensations (Borg scale). Prevention of hypoglycemia (self-measurement before and several hours after physical activity). Adapt the medication intake to the effort (intensity and duration).
Calcium channel blockers, ACE inhibitors, angiotensin II receptor blockers	Alter the ability to assess exercise intensity by heart rate. Risk of post-exercise hypotension.	Evaluation of exercise intensity by perceived sensations, especially respiratory sensations (Borg scale). Adapt the medication intake to the effort (intensity and duration).
Insulin and insulinotropic drugs	Hypoglycemic risk	Endocrinological advice if necessary. Prevention of hypoglycemia (self-measurement before and several hours after the physical activity)
Diuretics	Risk of dehydration, electrolyte disorders and therefore arrhythmia during exercise, especially in hot weather	Prevention of dehydration (with oral rehydration solutions)
Anticoagulants and antiplatelet drugs	Risk of hematomas and hemorrhages	Contra-indication to the practice of physical activity with impacts, risk of shocks and trauma.

Table 3.
Drug treatments and physical activity.

3.6.2 Specific indications according to the patient's pathologies

As explained above, the general indications for frequency, intensity, type and time are suitable for all patients regardless of their medical conditions as long as the recommendations for physical activity intensity and the pain threshold are met.

For all pathologies, the prescription must therefore include the general indications:

- Frequency: 150 to 300 min per week
- Moderate maximum intensity (Borg 4 to 5)
- Work on strengthening, coordination, balance and flexibility
- Respect for the rule of no pain

These general indications can be supplemented by specific indications considering the pathologies from which the patient suffers. The main specific indications are summarized in **Table 4**.

High blood pressure	Combined physical activities with endurance and strength exercises with a focus on aerobic endurance Moderate intensity (Borg scale 4 to 5) Precautions related to antihypertensive treatments
Coronary artery diseases	Combined physical activities with endurance and strength exercises Moderate intensity (Borg scale 4 to 5) Precautions related to antihypertensive and anticoagulants treatments Knowing how to spot the signs of myocardial ischemia
Cerebrovascular accident and transient ischemic attack	Combined physical activities with endurance and strength exercises Moderate intensity (Borg scale 4 to 5) Balance and coordination exercises Importance of daily physical activities in addition to supervised physical activities Precautions related to anticoagulants treatments
Chronic heart failure	Combined physical activities with endurance and strength exercises with a focus on strength exercises Low to moderate intensity (Borg scale 1 to 5) Contraindication of high intensities No physical activity in poor environmental conditions (cold, heat, pollution)
Type 2 diabetes	Combined physical activities with endurance and strength exercises Moderate intensity (Borg scale 4 to 5) Knowing how to spot the signs of myocardial ischemia Prevention of hypoglycemia if hypoglycemic treatments Quality of footwear
Chronic obstructive pulmonary disease	Combined physical activities with an emphasis on strength exercises Low to moderate intensity (Borg 1 to 4) Breathing exercises SaO_2 monitoring and cessation of physical activity when <90% Blood pressure monitoring No physical activities in poor environmental conditions (cold, heat, pollution) Combined with respiratory rehabilitation by physiotherapy when necessary

Asthma	Combined physical activities Moderate intensity (Borg 4 to 5) Have your bronchodilator treatment during physical activity No physical activities in poor environmental conditions (cold, heat, pollution)
Cancers	Combined physical activities Intensity: gradually reaching intensities "sustained" (Borg 5 to 7) taking into account cardiovascular limitations Prescribe physical activities before, during, and after chemotherapy treatment
Obesity	Combined physical activities starting at 150 minutes per week to reach gradually 300 minutes per week Moderate and high intensity (Borg 4 to 7 depending on cardiovascular risk) Respect the principle of joint sparing (physical activities in discharge: swimming, aquagym, velo...) Combine with interval training taking into account cardiovascular limitations
Multiple sclerosis	Combined physical activities Low to moderate intensity (Borg 1 to 5) Stretching and Balance Exercises Comply with the maximum symptomatic time Limit the increase in body temperature during physical activities and limit physical activities in hot weather Limiting physical activities during active flare-ups Combine with physiotherapy rehabilitation when necessary
Depression	Combined physical activities Moderate and high intensity (Borg 4 to 7 depending on cardiovascular risk) Type marche, yoga, qi gong, etc.
Parkinson's disease	Combined physical activities with a focus on endurance exercises Moderate intensity (Borg 4 to 5) Balance and coordination exercises Walking, yoga, dance type Combine with physiotherapy rehabilitation when necessary
Aging	Combined physical activities with an emphasis on strength exercises (especially of the lower limbs) Moderate intensity (Borg 4 to 5) Flexibility, balance and coordination exercises Memory exercises Secure risky movements (rapid movements, stairs, steps, rotation and knee flexion) Proper hydration before, during and after physical activities Combine with physiotherapy rehabilitation when necessary
Osteoporosis	Favor weight-bearing exercises such as walking and limit micro-gravity exercises such as water aerobics Secure risky movements (rapid movements, stairs, steps, rotation and knee flexion) Contraindication to the practice of physical activities with impacts, risk of shock and trauma
Fibromyalgia, myalgic encephalomyelitis / chronic fatigue syndrome, long covid	Combined physical activity with a focus on aerobic endurance Low to moderate intensity (Borg 2 to 5) Stretching and flexibility exercises Type: walking, cycling, dancing, water activity, tai chi Respect the maximal symptomatic time and the rule of no pain Combine with physiotherapy rehabilitation when necessary

Chronic inflammatory rheumatic diseases	Combined physical activities with a focus on strength exercises Moderate intensity (Borg 4 to 5) Stretching, flexibility and balance exercises Respect the rule of no pain and the principle of joint sparing Limit or even stop physical activities during inflammatory phases Footwear quality Combine with physiotherapy rehabilitation when necessary
Osteoarthritis and tendinopathy	Combined physical activities with a focus on strength, flexibility exercises and on proprioception, especially for the affected joints Favor passive amplitude work (when possible) and isometric reinforcement work (limit repetitions) Limit or even stop physical activities during congestive flare-ups Respect the rule of no pain and the principle of joint sparing Contraindication to the practice of physical activities with impacts, risk of shock and trauma Combine with physiotherapy rehabilitation when necessary

Table 4.
Specific indications according to pathologies.

In certain pathologies and health conditions, the notion of maximum symptomatic time (MST) must be considered and explained to the patient [22]. This is the time after which an activity causes pain to appear or exacerbate. To calibrate the duration of physical activity sessions, the patient must remain at 50 to 70% of his MST. He must therefore break up the sequences of activities with phases of rest, stretching and relaxation. This notion is important in multiple sclerosis, fibromyalgia, myalgic encephalomyelitis and long Covid, but also in people with fragile physical conditions related to age, cancer, heart or respiratory failure.

4. Conclusion

The systematic approach proposed in this chapter makes it possible to integrate the prescription of physical activity into the therapeutic management of patients suffering from chronic diseases and/or with risk factors. Followed systematically, the six steps proposed make it possible to eliminate the rare contraindications, most often temporary, requiring a specific medical check-up and/or a specialized advice and to specify the general and specific indications for adapting and securing the practice of physical activity.

The medical prescription of physical activity has several essential objectives [23]. It defines a safe and appropriate framework by specifying the general indications—intensity, frequency, type, duration—as well as specific indications and precautions. It also participates in the therapeutic education of the patient by informing him or her of advice for good practice (the 10 golden rules, respect for the level of intensity and the pain threshold, the warm-up and recovery phases, the precautions according to each pathology) and by accompanying him or her toward a more active lifestyle [24]. Finally, the medical prescription is an important factor in the patient's motivation and adherence to physical activity. It thus contributes to the regularity and sustainability of the practice, which are key elements for an effect on health. The medical prescription of physical activity must therefore become a central element of the therapeutic arsenal of doctors to meet the challenge of chronic diseases and the maintenance of autonomy.

Author details

Eugénie d'Alessandro[1,2]

1 Department of Public Health, Service communal d'hygiène et de santé, Salon de Provence, France

2 Centre Norbert Elias, École des Hautes Études en Sciences Sociales, Marseille, France

*Address all correspondence to: eugeniealessandro@hotmail.com

References

[1] Physical Activity Guidelines Advisory Committee. Physical Activity Guidelines Advisory Committee Scientific Report. Washington (DC): U.S. Department of Health and Human Services; 2018

[2] Dai J, Menhas R. Sustainable development goals, sports and physical activity: The localization of health-related sustainable development goals through sports in China: A narrative review. Risk Management and Health Policy. 2020;**13**:1419-1430

[3] INSERM. Activité Physique, Prévention et Traitement des Maladies Chroniques, Synthèse et Recommandations. France: Editions EDP Sciences; 2019

[4] Masanovic B et al. Editorial, physical activity as a natural cure for non-communicable diseases, front. Public Health. 2023;**11**:1209569

[5] World Health Organization. Global Levels of Physical Inactivity in Adults, Off Track for 2030. Geneva; 2024

[6] Hau Chong K, Suesse T, Cross P, et al. Pooled analysis of physical activity, sedentary behavior, and sleep among children from 33 countries. JAMA Pediatrics. 2024;**178**(11):1199-1207

[7] Strain T et al. National, regional, and global trends in insufficient physical activity among adults from 2000 to 2022: A pooled analysis of 507 population-based surveys with 5.7 million participants, lancet. Globalization and Health. 2024;**12**:e1232-e1243

[8] Jefferis B, Fagerland MW, et al. Dose-response associations between accelerometry measured physical activity and sedentarity time and all-cause mortality: Systematic review and harmonized meta-analysis. BMJ. 2019;**366**:14570

[9] Zhao R et al. The dose-response associations of sedentary time with chronic diseases and the risk for all-cause mortality affected by different health status: A systematic review and meta-analysis. The Journal of Nutrition, Health & Aging. 2020;**24**(1):63-70

[10] Gilchrist S et al. Association of sedentary behavior with cancer mortality in middle-aged and older US adults. JAMA Oncology. 2020;**6**(8):1210-1217

[11] World Health Organization. Global Recommendations on Physical Activity for Health. Geneva; 2010

[12] World Health Organization. Guidelines on Physical Activity and Sedentary Behavior. Geneva; 2020

[13] Larry Dustine J et al. Chronic diseases and the link to physical activity. Journal of Sport and Health Science. 2013;**2**:3-11

[14] Anderson E, Larry Dustin J. Physical activity, exercise and chronic diseases: A brief review. Sports Medicine and Health Sciences. 2019;**1**:3-10

[15] ACSM's. Guidelines for Exercise Testing and Prescription. 11th ed. USA: Wolters Kluwer; 2021

[16] Depiesse F, Coste O. Prescription des activités physiques en prévention et en thérapeutique. 2° ed. France: Elsevier Masson; 2022

[17] Haute Autorité de Santé, Guide. Consultation et prescription médicale

d'activité physique à des fins de santé chez l'adulte. 2022

[18] The physical activity readiness questionnaire for everyone. Available from: https://eparmedx.com/

[19] Score Charts. Available from: https://www.heartscore.org/en_GB/

[20] Williams N. The Borg rating of perceived exertion scale. Occupational Medicine. 2017;**65**(5):404-405

[21] French Federation of Cardiology. The Ten Golden Rules. Available from: https://www.fedecardio.org/je-m-informe/les-10-regles-d-or-du-sportif/

[22] Ranque Garnier S et al. Fibromyactiv: Single-center, prospective, randomized pilot-study. Adapted physical exercise efficacy on quality of life of patients with fibromyalgia. Douleurs. 2017;**18**(2):87-104

[23] Thornton J et al. Physical activity prescription: A critical opportunity to address a modifiable risk factor for the prevention and management of chronic disease: A position statement by the Canadian academy of sport and exercise medicine. British Journal of Sports Medicine. 2016;**50**:1109-1114

[24] Gasparini W, Knobé S, Didierjan R. Physical activity on medical prescription: A qualitative study of factors influencing take-up and adherence in chronically ill patients. Health Education Journal. 2015;**74**(6):720-731

Chapter 11

Physical Activity as a Mediator between Sleep, Mental Health, and Well-Being in Active Individuals

Monoem Haddad

Abstract

The relationship between physical activity, sleep quality, and mental health is multifaceted and vital for maintaining overall well-being. Regular physical activity plays a crucial role in enhancing psychological health, improving sleep, and increasing life satisfaction. Exercise has been consistently linked to reduced symptoms of depression, anxiety, and stress while regulating sleep patterns. These effects occur through complex neurophysiological mechanisms involving neurotransmitter regulation, hormonal balance, and cognitive improvements. This chapter explores the mediating role of physical activity in the relationship between sleep and mental health, emphasizing its importance in promoting better well-being. The chapter highlights evidence from various studies demonstrating that physical activity can improve sleep quality, reduce mental health symptoms, and prevent the onset of psychological disorders. Special attention is given to specific populations such as athletes, students, and older adults, considering their unique needs and benefits from physical activity. It concludes with practical strategies for incorporating exercise into daily routines to break the cycle of poor sleep and mental health, offering a cost-effective, non-pharmacological intervention that supports both mental and physical health.

Keywords: physical activity, sleep quality, mental health, well-being, exercise, depression, anxiety, sleep disorders, physical health, preventive care

1. Introduction: The triangular relationship

The relationship between physical activity, sleep quality, mental health, and overall well-being is complex, yet critically important for maintaining optimal health. These elements are deeply interlinked, influencing each other in multifaceted ways. While physical activity is often primarily associated with its benefits on physical health, its impact extends far beyond this, playing a significant role in enhancing psychological well-being, sleep quality, and overall life satisfaction. Regular exercise has been consistently shown to reduce symptoms of mental health disorders such as anxiety, depression, and stress [1, 2], while also improving sleep patterns by regulating sleep duration, reducing sleep latency, and enhancing sleep quality [3, 4]. This chapter explores how physical activity serves as a mediator between sleep and mental health, highlighting its importance in fostering better well-being in active individuals.

Physical activity not only contributes to physiological health but also significantly influences psychological functioning. The role of exercise as a mediator in the relationship between sleep and mental health has become an important area of research. A mediator variable is one that explains the process through which one factor influences another, and in this context, physical activity is the key mechanism that improves sleep, which in turn reduces mental health issues such as anxiety and depression [4, 5]. Regular engagement in physical activity stimulates neurotransmitters such as serotonin and dopamine, which are essential for regulating mood, enhancing emotional resilience, and improving cognitive function [6]. These physiological changes, along with the psychological benefits of exercise, create a positive feedback loop where improvements in sleep enhance mental health and vice versa [7]. By acting as a mediator, physical activity helps to break the cycle of poor sleep and mental health, offering a pathway to improved well-being for individuals engaged in regular exercise.

The bidirectional nature of the relationship between sleep and mental health has been well documented. Sleep disturbances often lead to the onset of mental health issues such as depression, anxiety, and stress, while mental health challenges can worsen sleep patterns, creating a vicious cycle that is difficult to overcome [5, 8]. This interdependence underscores the importance of improving sleep as a means to prevent or alleviate mental health disorders. In this sense, physical activity emerges as a critical tool for breaking this cycle. Exercise serves as a non-pharmacological intervention that can address both sleep disturbances and mental health issues simultaneously [5]. Research has consistently shown that physical activity not only improves sleep but also provides significant relief from mental health conditions, positioning it as an essential component of preventive medicine [2].

The growing evidence supporting the mental health and sleep benefits of physical activity highlights its potential as a preventive measure. Regular exercise, when integrated into daily routines, offers a simple, cost-effective solution to managing and preventing sleep disturbances and mental health issues [1]. This is particularly relevant in the context of current healthcare systems, where there is an increasing shift toward preventive care strategies that emphasize lifestyle modifications over pharmacological treatments [4]. Incorporating physical activity as a foundational element of preventive medicine is essential for promoting better psychosocial health and reducing the prevalence of sleep and mental health disorders globally [5, 6].

Given the clear benefits of physical activity on sleep and mental health, this chapter will explore the underlying mechanisms by which exercise influences these areas. It will review both the physiological and psychological pathways through which physical activity improves sleep and mental health, as well as provide evidence-based recommendations for incorporating exercise into the daily lives of active individuals. Special attention will be paid to specific populations such as athletes and students, who may experience unique benefits from regular physical activity, as well as to practical strategies for overcoming barriers to exercise that often hinder its implementation [5].

2. Literature review: The mediating role of physical activity

The connection between physical activity, sleep quality, and mental health is increasingly recognized as an essential area of research in the fields of psychology, exercise science, and public health. Studies have shown that regular physical activity can enhance sleep quality and improve mental health outcomes, such as reducing anxiety, depression, and stress [9, 10]. However, the underlying mechanisms that explain how physical activity influences both sleep and mental health are complex. One prominent explanation is the mediating role of physical activity in improving sleep and, consequently, mental health, suggesting that exercise can indirectly influence psychological well-being by improving sleep quality [1, 3].

Recent studies have highlighted the discrepancies between subjective and objective sleep assessments, which complicate the interpretation of the impact of physical activity on sleep and mental health. Haddad et al. [11] investigated these discrepancies in university students and found that self-reported sleep duration and quality were often inaccurate compared to objective measures like actigraphy. This misalignment suggests that previous studies relying solely on subjective sleep assessments may have underestimated or misrepresented the true impact of physical activity on sleep and mental health. Thus, integrating both subjective and objective measures of sleep is crucial for more accurate findings in this area.

Physical activity is commonly seen as a mediator that affects the sleep-mental health connection in multiple ways. For example, physical activity can improve sleep by regulating the circadian rhythm, promoting relaxation, and reducing anxiety [12]. In turn, better sleep quality has been shown to reduce the risk of developing mental health issues such as depression and anxiety [13]. Conversely, sleep deprivation and poor-quality sleep are significant risk factors for mental health disorders, making sleep a critical factor in understanding the impact of physical activity on mental health [14].

2.1 Physical activity and sleep

Numerous studies have documented the positive effects of physical activity on sleep quality. Exercise is known to improve sleep duration, sleep latency (the time taken to fall asleep), and overall sleep quality [15]. These effects are likely due to various physiological processes that are stimulated by physical activity. For example, regular exercise can enhance the function of the body's thermoregulatory system, leading to a decrease in body temperature post-exercise, which can promote faster sleep onset and more restful sleep [16]. Furthermore, physical activity has been shown to regulate the circadian rhythm, a biological clock that governs sleep–wake cycles, by aligning it with external cues such as sunlight exposure and activity levels [17].

The type and intensity of exercise also influence the sleep outcomes. Aerobic exercise, such as walking, running, or cycling, has been found to be particularly beneficial for improving sleep in both young and older adults [18]. On the other hand, resistance training and strength exercises can also have positive effects on sleep, though the benefits may be more pronounced when combined with aerobic exercise [15]. Moderate to vigorous-intensity exercise tends to produce better sleep outcomes compared to light-intensity activity [19].

A meta-analysis by Kredlow et al. [20] further supports the claim that physical activity significantly improves sleep, with effects being observed across different populations, including individuals with insomnia, depression, and anxiety disorders. It is important to note that the timing of exercise also plays a role in sleep outcomes. Exercising too close to bedtime can potentially disrupt sleep by raising core body temperature and stimulating the nervous system, while exercising earlier in the day has been associated with better sleep quality [15].

2.2 Physical activity and mental health

Beyond its impact on sleep, physical activity has been extensively studied for its effects on mental health. Regular exercise is known to reduce symptoms of depression, anxiety, and stress, with the therapeutic effects often being comparable to traditional treatments such as psychotherapy and pharmacological interventions [12]. The effects of exercise on mental health are mediated by several physiological mechanisms, including the release of endorphins and neurotransmitters such as serotonin and dopamine, which are known to regulate mood and emotional responses [20].

One of the key mechanisms through which physical activity influences mental health is by reducing levels of cortisol, a hormone associated with stress. Regular physical activity helps to regulate cortisol levels, preventing the negative impact of chronic stress on both mental health and sleep [21]. Moreover, exercise can improve self-esteem and self-efficacy, which contribute to better psychological well-being and a more positive outlook on life [22]. Studies have also shown that physical activity can enhance cognitive function, reduce rumination, and improve emotional regulation, all of which are beneficial for managing stress and anxiety [23].

The relationship between physical activity and mental health is complex and can vary depending on factors such as the individual's baseline mental health status, the type of exercise performed, and the frequency and duration of activity. For example, long-term, consistent physical activity is more effective at reducing depression and anxiety compared to short-term bouts of exercise [12]. Additionally, group-based physical activities, such as team sports or exercise classes, may provide additional social benefits that further enhance mental health outcomes [24].

2.3 Mediating role of physical activity between sleep and mental health

As outlined previously, physical activity serves as a mediator between sleep and mental health. Research has shown that exercise not only improves sleep but also mitigates the negative effects of poor sleep on mental health [13]. For instance, a study by Sharma [19] found that individuals with poor sleep who engaged in regular physical activity showed significant improvements in both sleep quality and mental health, compared to those who did not exercise. This is supported by other research suggesting that exercise-induced improvements in sleep can reduce anxiety and depressive symptoms [12]. Additionally, Haddad et al. [25] found that physical activity also has a significant impact on cognitive functioning, which in turn mediates both sleep and mental health outcomes. Their study with university students revealed that regular physical activity not only enhanced sleep quality but also improved cognitive performance, which played a role in reducing stress and anxiety, thus enhancing overall mental health. This highlights the critical role of exercise in not only improving sleep and mental health but also cognitive function, which contributes to an individual's emotional resilience and coping mechanisms.

The impact of physical activity on mental health can be further understood through the lens of learning environments. Haddad et al. [26] also investigated the effects of online versus face-to-face learning environments on sleep, physical activity, and cognitive functioning among physical education students. They found that while online learning did not significantly impact sleep quality, it led to reduced physical activity, which is a critical factor in both sleep quality and mental health. These findings underline the importance of considering environmental factors, such as the structure of learning environments, in future research on physical activity and mental health.

The mechanisms that mediate this relationship are likely a combination of physiological and psychological processes. On the physiological side, physical activity improves sleep through its effects on the autonomic nervous system, hormone regulation, and neurotransmitter activity, which collectively enhance mood and reduce stress [12]. Psychologically, physical activity can improve mood, reduce stress, and enhance self-esteem, all of which contribute to better mental health outcomes [23]. Thus, exercise provides a pathway to improving both sleep and mental health, reinforcing the idea that physical activity is not only beneficial for physical health but also plays a critical role in overall well-being.

2.4 The need for further research

While the existing literature provides compelling evidence that physical activity plays a crucial role in mediating the relationship between sleep and mental health, there are still gaps that require further exploration. Future research should focus on the mechanisms by which exercise influences sleep and mental health, with a particular emphasis on understanding how different types of physical activity, exercise intensity, and timing affect these outcomes. Additionally, more research is needed to determine how physical activity interventions can be personalized for individuals with specific sleep disorders or mental health conditions.

Furthermore, longitudinal studies are required to better understand the long-term effects of physical activity on sleep and mental health. Most of the existing studies are cross-sectional or short-term in nature, and more evidence is needed to determine the sustainability of the benefits of exercise on sleep and mental health over time. Studies examining the effects of exercise on vulnerable populations, such as individuals with chronic sleep disorders, severe mental health conditions, or those living in high-stress environments, are also necessary to further refine exercise-based interventions for improving sleep and mental health.

3. Theoretical framework: Mechanisms of mediation

The relationship between physical activity, sleep, and mental health is complex, with multiple mechanisms through which physical activity influences both sleep quality and mental well-being. These mechanisms can be broadly categorized into physiological processes involving neurochemicals, hormones, and circadian rhythms and psychological processes such as self-esteem, cognitive function, and social interaction. In this section, we explore these mechanisms in greater detail.

3.1 Physiological mechanisms

One of the most significant physiological mechanisms through which physical activity mediates the effects of sleep on mental health is the regulation of neurochemical and hormonal systems. Physical activity induces changes in neurotransmitter levels, which affect mood regulation and emotional well-being. Notably, physical activity increases the production of serotonin, dopamine, and gamma-aminobutyric acid (GABA), which are known to play essential roles in improving mood, reducing anxiety, and promoting relaxation [23]. By stimulating the production of these neurotransmitters, exercise helps to regulate emotional responses and reduce feelings of depression and anxiety [23].

3.1.1 Serotonin and dopamine

Exercise enhances serotonin availability, which is crucial for maintaining a positive mood and reducing symptoms of depression. Studies have shown that aerobic exercise, in particular, increases serotonin production in the brain, contributing to improved mood and emotional stability [22]. Similarly, dopamine, which is involved in reward and motivation pathways, is also increased following exercise. This increase in dopamine levels helps to improve mood and motivation, making exercise an effective strategy for managing mental health issues [27].

3.1.2 Endorphins and cortisol regulation

Endorphins, which are released during physical activity, play a significant role in improving mood and alleviating stress. These endogenous chemicals act as natural painkillers and mood enhancers, providing feelings of euphoria and well-being. Additionally, physical activity regulates cortisol levels, a hormone closely linked to stress. Chronic high cortisol levels are associated with both mental health disorders and sleep disturbances, but regular exercise helps to lower cortisol levels, preventing the negative effects of stress on sleep and mental health [4]. By regulating cortisol and boosting endorphin levels, physical activity helps reduce the physiological effects of stress, promoting better sleep and mental health.

3.1.3 Circadian rhythm and sleep regulation

Physical activity also plays a vital role in regulating the circadian rhythm, the body's internal clock that controls sleep–wake cycles. Exercise has been shown to enhance the synchronization of the circadian rhythm, leading to better sleep quality and more restful sleep [28]. This regulation of sleep–wake cycles is particularly beneficial for individuals with insomnia and other sleep disorders. Regular exercise helps to align the body's internal clock with external cues, such as light exposure and physical activity, thereby improving sleep patterns. The quality of sleep is a significant determinant of mental health, and by improving sleep quality, physical activity indirectly improves mental health outcomes.

3.2 Psychological mechanisms

In addition to the physiological effects, psychological mechanisms also play a crucial role in how physical activity mediates the relationship between sleep and mental health. One of the most prominent psychological mechanisms is the improvement of self-esteem and self-efficacy. Regular physical activity, particularly when it involves goal setting and achievement, boosts self-esteem by improving body image, fitness levels, and overall physical health. This increase in self-esteem leads to better psychological well-being and a more positive outlook on life [23].

3.2.1 Self-esteem and self-efficacy

Physical activity has been consistently linked to improvements in self-esteem and self-efficacy, which are important psychological factors that influence mental health. Self-esteem refers to an individual's sense of worth, and self-efficacy refers to their belief in their ability to succeed in specific tasks. By achieving exercise-related goals and improving physical fitness, individuals experience greater confidence and control, which can reduce feelings of helplessness associated with mental health issues like depression and anxiety [12].

3.2.2 Cognitive function and emotional regulation

Exercise is also associated with improvements in cognitive function, including attention, memory, and executive function. Cognitive improvements help individuals manage stress, anxiety, and depression by enhancing emotional regulation and coping skills [27]. Regular physical activity promotes neuroplasticity—the brain's ability to adapt and reorganize itself—which leads to improved cognitive function and mental flexibility [24]. Furthermore, physical activity has been shown to reduce rumination, a negative thinking pattern commonly associated with depression and anxiety. By disrupting the cycle of rumination, exercise helps individuals focus on positive thoughts and engage more fully in life, further enhancing mental health [29].

3.2.3 Social interaction and support

Group-based physical activities, such as team sports or exercise classes, provide additional social benefits that further enhance mental health outcomes. Social interactions are crucial for mental well-being, as positive relationships are associated with lower levels of depression, anxiety, and stress. Group exercise provides opportunities for individuals to connect with others, receive support, and experience a sense of belonging, all of which contribute to better mental health [22]. Social support is a key factor in improving emotional regulation and coping with stress, which further mediates the positive effects of physical activity on mental health.

3.3 Mediation through behavioral activation

Another psychological mechanism by which physical activity influences mental health is through behavioral activation. Behavioral activation is a therapeutic strategy used to treat depression, encouraging individuals to engage in rewarding and fulfilling activities. Physical activity serves as an effective form of behavioral activation by promoting positive emotions, enhancing mood, and reducing feelings of sadness and isolation. The structured nature of exercise routines provides individuals with a sense of purpose and accomplishment, which is particularly beneficial for those struggling with depression and anxiety [30]. By increasing engagement in positive behaviors, physical activity helps break the cycle of negative thoughts and emotions, fostering a healthier mental state.

4. Case studies and research evidence: Real-world examples

The impact of physical activity on sleep and mental health is not only a theoretical concept but has also been demonstrated in numerous real-world studies and case examples. In this section, we explore a variety of research studies and case studies to illustrate the practical application of the mediation mechanisms discussed earlier.

4.1 Case study 1: Physical activity and mental health in clinical populations

One of the most well-known examples of the benefits of physical activity on mental health comes from clinical populations, particularly individuals suffering from depression and anxiety. Research has shown that physical activity can be as effective as pharmacotherapy for treating depression, with the added benefit of fewer side effects.

A study by Babyak et al. [2] investigated the long-term effects of exercise as a treatment for major depressive disorder (MDD). Participants in this randomized controlled trial were divided into three groups: an exercise group, a medication group (using antidepressants), and a combined exercise and medication group. The results showed that after 16 weeks, all groups experienced significant reductions in depressive symptoms. However, the exercise group maintained its therapeutic benefits even after 10 months, suggesting that exercise could be a viable long-term strategy for managing depression without the need for medication [2].

Similarly, Blumenthal et al. [10] examined the combination of exercise and pharmacotherapy for treating MDD. The study revealed that patients who engaged in regular physical activity experienced improvements in both their physical health and their mental health, with many patients reporting improved sleep, reduced anxiety, and better overall well-being. These findings underscore the importance of integrating physical activity into treatment plans for patients with depression and anxiety disorders.

4.2 Case study 2: Exercise and sleep in university students

University students often experience disrupted sleep patterns and increased mental health challenges due to academic stress, social pressures, and irregular sleep habits. A study by De Nys et al. [4] investigated the effects of physical activity on sleep quality in university students. The results showed that students who engaged in regular moderate to vigorous physical activity reported improved sleep quality, increased sleep duration, and a decrease in sleep latency compared to students who were sedentary.

The positive impact of physical activity on sleep also had indirect benefits for mental health. Students who reported better sleep were also less likely to experience symptoms of anxiety and depression. This finding highlights the potential of physical activity as a dual intervention for improving both sleep and mental health in this population. Additionally, group-based physical activities, such as intramural sports or fitness classes, were found to be particularly effective in improving social connections and reducing feelings of isolation, which are common contributors to poor mental health among students.

4.3 Case study 3: Physical activity and mental health in older adults

The benefits of physical activity for mental health and sleep are not limited to younger populations. In fact, older adults may experience even more significant improvements in both mental health and sleep as a result of regular physical activity. A study by Fank et al. [18] reviewed the impact of exercise on sleep in older adults. The review found that older adults who engaged in regular aerobic exercise experienced better sleep quality, reduced symptoms of insomnia, and decreased daytime sleepiness. These improvements in sleep were accompanied by reductions in symptoms of anxiety and depression.

In one specific case, an older adult woman with chronic insomnia and depression began participating in a community walking program. After three months of regular walking, her sleep quality improved, and her depressive symptoms significantly reduced. The improvements in sleep were not only beneficial for her mental health but also improved her cognitive function, allowing her to engage more fully in social and daily activities.

4.4 Research evidence: The role of physical activity in preventing mental health issues

In addition to case studies, research has consistently demonstrated the role of physical activity in preventing mental health disorders, particularly in populations at risk. A study by Singh et al. [12] reviewed the effectiveness of physical activity interventions in improving depression, anxiety, and distress. The research found that individuals who engaged in regular physical activity were significantly less likely to develop depression or anxiety disorders over time. Furthermore, interventions targeting both physical activity and sleep were found to be particularly effective in preventing the onset of these mental health conditions.

The research also highlighted that different types of physical activity—such as aerobic exercise, strength training, and mind–body exercises (e.g., yoga)—can have different effects on mental health. While aerobic exercises are effective in improving mood and reducing stress, strength training was shown to improve self-esteem and body image, which are key factors in managing depression and anxiety. Mind–body exercises, such as yoga, were particularly effective in reducing rumination and improving emotional regulation, both of which are associated with lower levels of anxiety and depression.

4.5 Research evidence: Group-based physical activities and social benefits

Another important factor in the relationship between physical activity, sleep, and mental health is the social support provided by group-based physical activities. Biddle and Asare [22] reviewed the mental health benefits of physical activity for children and adolescents, emphasizing that team sports and exercise classes promote not only physical health but also social well-being. Participating in group-based physical activities helps individuals develop social connections, reduce feelings of loneliness, and provide a sense of belonging—all of which are crucial for mental health.

A specific study by Biddle et al. [9] looked at the impact of group-based physical activity interventions on mental health in adolescents. The results indicated that adolescents who participated in group-based physical activities reported lower levels of depression, anxiety, and distress. These findings suggest that the social benefits of group-based exercise play a significant role in improving mental health outcomes, in addition to the direct benefits of physical activity on mood and well-being.

5. Practical implications: Designing physical activity programs

Designing effective physical activity programs requires a clear understanding of the needs and goals of different populations, including those seeking improvements in mental health, sleep, and overall well-being. The primary objective of these programs is to maximize the benefits of physical activity, not only for physical health but also for mental health, including enhancing sleep quality and reducing symptoms of depression, anxiety, and stress.

In this section, we will outline key factors that should be considered when designing physical activity programs and provide evidence from research that supports the implementation of these strategies.

5.1 Tailoring physical activity to specific populations

The design of physical activity programs should consider the specific needs, limitations, and goals of the target population. For instance, older adults, individuals with mental health disorders, and university students may have different physical activity needs and goals. Tailoring the intensity, frequency, and type of exercise to these populations is crucial for ensuring effectiveness.

5.1.1 Older adults

Programs designed for older adults should emphasize low-impact, moderate-intensity exercises to reduce the risk of injury. Research shows that older adults can benefit significantly from walking programs, strength training, and flexibility exercises. A study by Fank et al. [18] demonstrated that moderate aerobic exercise improved sleep quality and reduced symptoms of anxiety and depression in older adults.

5.1.2 University students

University students often experience high levels of stress, sleep disturbances, and mental health challenges. A study by De Nys et al. [4] showed that physical activity programs targeting university students improved sleep patterns and reduced stress levels. Programs focusing on both aerobic and resistance training exercises can help improve mood and cognitive function. Incorporating group activities, such as intramural sports or fitness classes, also provides social benefits that enhance mental well-being.

5.1.3 Individuals with mental health disorders

For individuals experiencing mental health issues, such as depression and anxiety, physical activity programs can be an effective non-pharmacological intervention. A meta-analysis by Schuch et al. [23] found that exercise is as effective as psychotherapy and medication for improving symptoms of depression. Programs should focus on increasing physical activity levels gradually, incorporating activities that the individuals enjoy to ensure long-term adherence.

5.2 Program duration and frequency

The frequency, intensity, and duration of physical activity programs play a crucial role in determining their effectiveness. For sleep and mental health improvements, research suggests that *moderate-intensity aerobic exercise* is optimal. The American College of Sports Medicine (ACSM) recommends that individuals engage in at least *150 minutes of moderate-intensity exercise per week*, spread across several days.

5.2.1 Duration

Programs should aim for *at least 30 minutes per session* for most individuals. For those with more severe mental health symptoms, shorter sessions of *10–15 minutes* may be beneficial and easier to adhere to in the beginning. Over time, the duration and intensity of the exercises can be increased gradually to help individuals experience greater improvements.

5.2.2 Frequency

Physical activity should be performed *3–5 times per week* for optimal effects on mental health and sleep quality. Consistency is key, as research shows that regular exercise can lead to significant long-term benefits. A study by Kredlow et al. [20] found that regular physical activity not only improved sleep quality but also helped reduce the occurrence of sleep disorders in adults.

5.3 Exercise types for mental health and sleep benefits

5.3.1 Aerobic exercise

Aerobic exercise, such as walking, jogging, cycling, or swimming, has been consistently shown to improve both sleep and mental health. Studies show that aerobic exercise enhances sleep quality by regulating the sleep–wake cycle and increasing the amount of deep sleep. Additionally, it stimulates the release of endorphins, which help alleviate symptoms of anxiety and depression [27].

5.3.2 Resistance training

Incorporating resistance training exercises, such as weightlifting or bodyweight exercises, has been shown to improve self-esteem and reduce depressive symptoms. A study by Schuch et al. [23] emphasized that resistance training could be particularly effective in individuals with depression, enhancing mood and building muscle strength.

5.3.3 Mind-body exercises

Mind–body exercises such as yoga, Pilates, and tai chi can improve mental health and sleep through relaxation and stress reduction techniques. These exercises focus on breathing, flexibility, and mindfulness, which have been shown to reduce symptoms of anxiety and depression while enhancing sleep quality [24].

5.4 Group-based activities and social interaction

Group-based physical activities, such as team sports, fitness classes, or group walks, provide not only physical and mental health benefits but also social support, which plays a vital role in mental well-being. Participating in group-based activities helps reduce isolation, builds social networks, and enhances feelings of belonging and connectedness.

A study by Biddle & Asare [22] showed that adolescents who engaged in group-based physical activities had lower levels of anxiety and depression, highlighting the importance of social interactions in mental health outcomes. Group-based activities also improve adherence to exercise programs, as participants feel motivated by their peers and are more likely to remain engaged over time.

5.5 Barriers to exercise and strategies for overcoming them

While physical activity has significant benefits, several barriers may prevent individuals from engaging in regular exercise. Common barriers include lack of time, motivation, and access to facilities, as well as physical limitations due to age or injury. To overcome these barriers, it is important to incorporate strategies that make exercise more accessible, enjoyable, and sustainable.

5.5.1 Motivational strategies

Goal-setting, social support, and positive reinforcement can help increase motivation and adherence to physical activity programs. Encouraging individuals to set realistic, attainable goals and tracking progress can foster a sense of accomplishment and help maintain motivation [9].

5.5.2 Accessible exercise options

Programs should consider providing exercise options that are easily accessible, such as home-based workouts or outdoor activities, particularly for those with limited access to gyms or fitness centers. Simple exercises like walking or bodyweight exercises can be done in a variety of settings and require minimal equipment.

5.5.3 Supportive environments

Creating supportive environments where exercise is encouraged and valued can help individuals stay committed to their physical activity goals. For example, promoting physical activity within schools, workplaces, and communities can foster a culture of health and well-being that makes exercise a routine part of daily life.

6. Challenges and barriers to implementation

While physical activity programs offer significant benefits for mental health and sleep quality, several barriers exist that may prevent their widespread implementation and participation. These barriers can be individual, environmental, or systemic in nature. Understanding and addressing these challenges is essential for promoting physical activity as a key strategy for improving mental and physical health.

6.1 Individual barriers to physical activity participation

Several factors at the individual level can prevent people from engaging in physical activity programs, even when they are aware of the benefits. These include:

6.1.1 Lack of motivation

Many individuals struggle to stay motivated to engage in regular exercise, especially when dealing with mental health challenges such as depression, anxiety, or stress.

Research shows that mental health issues can reduce the desire to engage in physical activity, creating a vicious cycle of inactivity and worsening mental health. Biddle and Asare [22] emphasized that effective interventions need to address intrinsic motivation by focusing on enjoyment, perceived competence, and autonomy. Similarly, Satariano and McAuley [31] found that addressing motivation in low-income populations, where barriers to exercise are heightened, is critical for program success.

6.1.2 Physical limitations

People with chronic conditions, disabilities, or age-related limitations may face difficulties in participating in certain forms of physical activity. Programs that are too demanding or not adaptable to an individual's specific physical condition can discourage participation. Tailoring programs to the individual's capabilities is crucial, as shown by Fank et al. [18], who noted that older adults with chronic illnesses benefit from low-impact, low-intensity exercises such as walking or tai chi. Additionally, Kredlow et al. [20] suggested that for individuals with sleep disorders or other health challenges, gradual and customized programs may be more successful.

6.1.3 Lack of time

Busy schedules, work commitments, and family responsibilities often lead to a lack of time for physical activity. This is especially true in populations like university students and working adults. Singh et al. [12] found that one of the main barriers to regular exercise in young adults was a perceived lack of time, which led to reduced participation in physical activity programs. Similarly, Penedo and Dahn [32] highlighted time constraints as a barrier in community-based programs, noting that shorter sessions or flexible times may help increase engagement.

6.2 Environmental barriers to physical activity

Environmental factors play a significant role in the accessibility of physical activity programs. Barriers at this level include:

6.2.1 Lack of access to facilities

In many communities, especially in lower-income areas, access to gyms, parks, and safe spaces for physical activity is limited. This can prevent people from engaging in regular exercise. Barton and Pretty [24] found that the availability of green spaces and safe walking areas is positively correlated with increased physical activity and improved mental health outcomes. Same authors [24] suggested that creating community-based exercise programs in areas with limited access to formal facilities could help overcome this barrier.

6.2.2 Socioeconomic barriers

Financial constraints may prevent people from participating in physical activity programs, particularly those that require paid memberships or access to specialized facilities. Offering low-cost or free exercise programs is essential to overcome this barrier. Programs that are community-based and utilize outdoor spaces can reduce financial barriers while increasing accessibility. According to Satariano and McAuley [31], older adults, particularly those from lower-income backgrounds, face significant barriers to physical activity, and community-based outreach programs that utilize public spaces or provide free access can help increase participation.

6.2.3 Climate and weather

Weather conditions, especially in regions with extreme temperatures, can discourage outdoor exercise. Programs that are adaptable to indoor spaces or can be done at home (e.g., virtual fitness classes and home workout routines) may help mitigate the impact of poor weather on exercise habits. Hirshkowitz et al. [17] discussed the importance of creating flexible exercise programs that can be done in various settings, considering environmental constraints. Similarly, Tcymbal et al. [33] noted that the built environment plays a key role in encouraging or limiting physical activity. Their systematic review highlighted the importance of urban planning that takes sex and gender into account, emphasizing that accessible and safe spaces for exercise are crucial for increasing physical activity levels. They suggested that urban environments should be designed to facilitate year-round access to physical activity opportunities for all individuals, especially in underserved areas.

6.3 Systemic barriers to implementation

At the systemic level, there are several structural challenges that can hinder the widespread implementation of physical activity programs:

6.3.1 Insufficient support from health systems

Health systems often focus more on treatment rather than prevention. Integrating physical activity into healthcare routines as a preventive measure is a challenge. As noted by Merlo and Vela [5], integrating physical activity programs into public health initiatives and healthcare services is essential for achieving population-wide improvements in mental health and sleep. Mammen and Faulkner [34] further emphasized the need for healthcare professionals to be better trained in promoting physical activity as part of a holistic approach to health, particularly given its significant benefits for mental health and overall well-being.

6.3.2 Lack of training for healthcare providers

Healthcare providers are often not trained to recommend or incorporate physical activity into their treatment plans. According to Schuch et al. [23], mental health professionals may not always be aware of the strong evidence supporting the benefits of physical activity for mental health, leading to missed opportunities for intervention. Brosse et al. [1] also found that while physical activity has mental health benefits, it is often underutilized due to a lack of healthcare provider training in this area.

6.3.3 Policy and infrastructure gaps

In many countries, public policies do not prioritize physical activity programs or infrastructure that supports exercise. For instance, urban planning may not include enough pedestrian-friendly streets, parks, or recreational facilities. Policymakers should focus on creating environments that encourage physical activity, such as safe walking paths, recreational centers, and integrated transport systems. Lee et al. [35] found that public health policies significantly influence the effectiveness of physical activity interventions, with well-designed policies helping to increase participation and adherence to exercise programs.

6.4 Strategies to overcome barriers

6.4.1 Increasing awareness and education

One of the most effective ways to overcome barriers to physical activity is through education and awareness campaigns. Informing individuals about the mental health and sleep benefits of physical activity can motivate them to get started and stay engaged. Public health campaigns should be designed to target both the general public and healthcare professionals. Marquez et al. [36] found that awareness campaigns focused on the long-term health benefits of physical activity helped increase participation rates in community-based programs.

6.4.2 Tailoring programs to individual needs

Personalized physical activity programs are more likely to succeed in helping individuals overcome barriers. This includes offering programs of varying intensities and durations, providing exercise options that suit individuals' health status, and incorporating activities that people enjoy. Programs should focus on building confidence and self-efficacy, which are key to long-term adherence. Wegner et al. [37] showed that personalized programs that take into account individual preferences and limitations are more likely to result in lasting behavioral change, especially in the context of exercise for anxiety and depression disorders.

6.4.3 Promoting social support

Social support is a critical factor in the success of physical activity programs. Group-based programs, such as community walking groups, fitness classes, or team sports, can provide social interaction, accountability, and motivation. Encouraging family and friends to participate in exercise programs together can also increase adherence. Trost et al. [38] found that social support played a significant role in increasing exercise participation among older adults. Bidzan-Bluma and Lipowska [39] further highlighted the importance of social networks in motivating people to continue exercising, especially in rural and underserved communities.

6.4.4 Policy changes and infrastructure development

Governments and organizations must prioritize the development of infrastructure that supports physical activity, such as building parks, bike lanes, and accessible public spaces. Public health policies should encourage exercise as part of healthcare prevention strategies and provide funding for community-based physical activity programs. Trost et al. [38] emphasized the need for government-led initiatives to improve access to physical activity resources, especially for vulnerable populations.

7. Future research directions

Despite the growing body of evidence highlighting the positive effects of physical activity on mental health and sleep, several gaps remain in our understanding of the mechanisms through which exercise influences well-being. Future research in this area should address these gaps by exploring several key areas:

7.1 Personalized exercise interventions

There is increasing recognition of the need for personalized exercise regimens that cater to the individual's physical health, sleep patterns, and mental health status. While standard recommendations such as moderate-intensity aerobic exercise are beneficial for many, future research should focus on developing customized interventions that consider the unique physical, psychological, and environmental factors affecting individuals. This could include integrating individual preferences, baseline mental health status, and barriers such as physical limitations or environmental factors into exercise plans. Personalized interventions could improve adherence and enhance long-term benefits, as research suggests that individualized programs are more likely to lead to sustained behavioral change [24].

7.2 Longitudinal studies on long-term benefits

Many existing studies have focused on short-term interventions and cross-sectional data. To fully understand the lasting effects of exercise on sleep and mental health, more longitudinal studies are required. These studies could track individuals over several years to evaluate whether the benefits of physical activity persist or even increase over time, and how they influence the long-term trajectory of mental health conditions such as depression and anxiety. Long-term research would also help clarify whether regular exercise can serve as a preventive tool, particularly in vulnerable populations [12, 23].

7.3 Exploring under-studied populations

While the effects of physical activity on sleep and mental health have been widely studied in the general population, research on specific groups such as older adults, individuals with chronic illnesses, and high-risk populations (e.g., people experiencing homelessness or those from low-income backgrounds) remains underexplored. These groups may face additional barriers to engaging in physical activity, and understanding how exercise can be tailored to meet their unique needs is essential. Recent research emphasizes the role of community-based programs in addressing these barriers and promoting exercise among underserved populations [34].

7.4 Mechanisms of action

Future studies should delve deeper into the specific mechanisms through which physical activity affects sleep and mental health. While there is evidence that physical activity can regulate neurotransmitters such as serotonin and dopamine, as well as cortisol levels, much remains unknown about the precise physiological and neurobiological pathways involved. For example, future studies could investigate the role of neuroplasticity and brain-derived neurotrophic factors (BDNF) in exercise-induced improvements in mental health and cognitive function [23]. Moreover, understanding the relationship between exercise, sleep quality, and inflammation could provide further insights into how exercise mitigates mental health conditions.

7.5 Technology and exercise monitoring

With advancements in technology, wearable devices and mobile apps are becoming valuable tools for monitoring physical activity and its effects on mental health. Future research should explore how these technologies can be integrated into exercise interventions to track progress, provide real-time feedback, and enhance motivation. Additionally, the use of virtual exercise programs could be explored, particularly in the context of reaching populations that face challenges such as geographical isolation or physical limitations. Virtual fitness classes or exercise apps could offer a flexible and accessible means of engaging in physical activity for individuals who are unable to attend in-person programs due to various barriers [38].

7.6 Global and cultural perspectives

Exercise and its benefits on mental health and sleep may be experienced differently across cultures and regions. Future research should examine how cultural factors influence the effectiveness of exercise interventions. For instance, while exercise is widely promoted in many Western countries, cultural differences in attitudes toward physical activity may affect participation rates and outcomes. Studies could explore the cultural determinants of exercise behaviors, identifying culturally appropriate strategies for engaging populations in different parts of the world. Additionally, research is needed to understand how environmental factors, such as air pollution or access to green spaces, interact with physical activity to affect mental health outcomes [24, 39].

7.7 Policy and environmental changes

While much attention has been focused on individual-level interventions, systemic and environmental changes are essential for creating an environment that supports physical activity. Research into public health policies that promote physical activity in urban planning, workplace settings, and educational institutions will help develop effective strategies for increasing participation in exercise programs. Investigating the role of government policy, particularly in addressing socioeconomic barriers, will be key to promoting physical activity as a mainstream preventive health measure [31].

8. Conclusion: Bringing it all together

The relationship between physical activity, sleep quality, and mental health is intricate and multifaceted. As evidence continues to mount, it is clear that physical activity serves as a powerful tool for improving both mental health and sleep quality, contributing significantly to overall well-being. Regular exercise has been shown to reduce symptoms of anxiety, depression, and stress, while also regulating sleep patterns, improving sleep quality, and reducing sleep disturbances. These effects are achieved through complex physiological and neurobiological mechanisms that involve neurotransmitter regulation, hormonal balance, and improved cognitive function.

However, while physical activity has clear benefits, there remains a need for more personalized interventions. Future research must explore the effectiveness of individualized exercise programs that account for physical, psychological, and social factors. This would ensure that exercise interventions can be tailored to meet the specific needs of diverse populations, including vulnerable groups such as older adults, those with chronic illnesses, and individuals from lower socioeconomic backgrounds. Moreover, there is a call for longitudinal studies to better understand the long-term effects of physical activity on mental health and sleep. These studies will provide valuable insights into how regular exercise can act as a preventive measure, not only improving quality of life but potentially reducing the incidence of mental health disorders.

Social support also plays a critical role in promoting and sustaining physical activity. Programs that foster community engagement and provide social networks are more likely to succeed, particularly for individuals who face barriers to exercise such as limited resources, physical limitations, or geographical isolation. Additionally, integrating technology, such as wearable devices and mobile health apps, can further enhance adherence and motivation to exercise. These tools can help track progress, offer real-time feedback, and provide personalized recommendations, making physical activity more accessible and engaging.

Public health policies and environmental changes are essential for creating a broader culture of physical activity. Government-led initiatives that focus on improving access to safe exercise spaces, creating supportive urban environments, and addressing socioeconomic barriers will be key to increasing participation rates. Research into the role of public health policies in promoting physical activity should be prioritized, particularly to support underserved communities that face additional challenges in accessing exercise resources.

In conclusion, while progress has been made in understanding the benefits of physical activity on sleep and mental health, future research is crucial in filling the gaps that remain. Personalized interventions, long-term studies, and a focus on underserved populations are necessary steps to maximize the positive impact of physical activity on well-being. By fostering a multidisciplinary approach that incorporates individual, social, and policy-level interventions, we can enhance the reach and effectiveness of exercise as a tool for improving mental health and sleep outcomes worldwide.

Acknowledgements

This work was supported by Qatar University under Collaborative Grants QUCG-CED-21-22-1 and QUCG-CED- 24/25-495.

Conflict of interest

The authors declare no conflict of interest.

Author details

Monoem Haddad
Sport Coaching Department, College of Sport Sciences, Qatar University, Doha, Qatar

*Address all correspondence to: mhaddad@qu.edu.qa

References

[1] Brosse AL, Sheets ES, Lett HS, Blumenthal JA. Exercise and the treatment of clinical depression in adults: Recent findings and future directions. Sports Medicine. 2002;**32**:741-760

[2] Babyak M, Blumenthal JA, Herman S, Khatri P, Doraiswamy M, Moore K, et al. Exercise treatment for major depression: Maintenance of therapeutic benefit at 10 months. Psychosomatic Medicine. 2000;**62**:633-638

[3] Driver HS, Taylor SR. Exercise and sleep. Sleep Medicine Reviews. 2000;**4**:387-402

[4] De Nys L, Anderson K, Ofosu EF, Ryde GC, Connelly J, Whittaker AC. The effects of physical activity on cortisol and sleep: A systematic review and meta-analysis. Psychoneuroendocrinology. 2022;**143**:105843

[5] Merlo G, Vela A. Mental health in lifestyle medicine: A call to action. American Journal of Lifestyle Medicine. 2022;**16**:7-20

[6] Belcher BR, Zink J, Azad A, CampbellCE,ChakravarttiSP,HertingMM. The roles of physical activity, exercise, and fitness in promoting resilience during adolescence: Effects on mental well-being and brain development. Biological Psychiatry: Cognitive Neuroscience and Neuroimaging. 2021;**6**:225-237

[7] Mahindru A, Patil P, Agrawal V. Role of physical activity on mental health and well-being: A review. Cureus. 2023;**15**:1-7

[8] Marino C, Andrade B, Campisi SC, Wong M, Zhao H, Jing X, et al. Association between disturbed sleep and depression in children and youths: A systematic review and meta-analysis of cohort studies. JAMA Network Open. 2021;**4**:e212373-e212373

[9] Biddle SJ, Vergeer I. Mental health benefits of physical activity for young people. In: The Routledge Handbook of Youth Physical Activity. London, UK: Routledge; 2020. pp. 121-147

[10] Blumenthal JA, Babyak MA, Doraiswamy PM, Watkins L, Hoffman BM, Barbour KA, et al. Exercise and pharmacotherapy in the treatment of major depressive disorder. Psychosomatic Medicine. 2007;**69**:587-596

[11] Haddad M, Abbes Z, Abdel-Salam A-SG. Bridging the gap: Investigating discrepancies between subjective and objective sleep assessment in university students. Biological Rhythm Research. 2025;**56**(1):63-74

[12] Singh B, Olds T, Curtis R, Dumuid D, Virgara R, Watson A, et al. Effectiveness of physical activity interventions for improving depression, anxiety and distress: An overview of systematic reviews. British Journal of Sports Medicine. 2023;**57**:1203-1209

[13] Zhao Y, Song J, Brytek-Matera A, Zhang H, He J. The relationships between sleep and mental and physical health of Chinese elderly: Exploring the mediating roles of diet and physical activity. Nutrients. 2021;**13**:1316

[14] Cohen S, Janicki-Deverts D, Miller GE. Psychological stress and disease. JAMA. 2007;**298**:1685-1687

[15] Dolezal BA, Neufeld EV, Boland DM, Martin JL, Cooper CB. Interrelationship between sleep and exercise: A systematic review. Advances in Preventive Medicine. 2017;**2017**:1364387

[16] Alnawwar MA, Alraddadi MI, Algethmi RA, Salem GA, Salem MA, Alharbi AA. The effect of physical activity on sleep quality and sleep disorder: A systematic review. Cureus. 2023;**15**:1-11

[17] Hirshkowitz M, Whiton K, Albert SM, Alessi C, Bruni O, DonCarlos L, et al. National Sleep Foundation's sleep time duration recommendations: Methodology and results summary. Sleep Health. 2015;**1**:40-43

[18] Fank F, da Silva Pereira F, Dos Santos L, De Mello MT, Mazo GZ. Effects of exercise on sleep in older adults: An overview of systematic reviews and meta-analyses. Journal of Aging and Physical Activity. 2022;**30**:1101-1117

[19] Ernest DK, Singh B, Shamna F, Onugha EA, Sharma SV. Unraveling the relationship between sleep behaviors and physical activity in the United States: Insights from 2015 to 2020 NHANES data. Journal of Activity, Sedentary and Sleep Behaviors. 2024;**3**:31

[20] Kredlow MA, Capozzoli MC, Hearon BA, Calkins AW, Otto MW. The effects of physical activity on sleep: A meta-analytic review. Journal of Behavioral Medicine. 2015;**38**:427-449

[21] Massar SA, Liu JC, Mohammad NB, Chee MW. Poor habitual sleep efficiency is associated with increased cardiovascular and cortisol stress reactivity in men. Psychoneuroendocrinology. 2017;**81**:151-156

[22] Biddle SJ, Asare M. Physical activity and mental health in children and adolescents: A review of reviews. British Journal of Sports Medicine. 2011;**45**:886-895

[23] Schuch FB, Vancampfort D, Richards J, Rosenbaum S, Ward PB, Stubbs B. Exercise as a treatment for depression: A meta-analysis adjusting for publication bias. Journal of Psychiatric Research. 2016;**77**:42-51

[24] Barton J, Pretty J. What is the best dose of nature and green exercise for improving mental health? A multi-study analysis. Environmental Science & Technology. 2010;**44**:3947-3955

[25] Haddad M, Abbes Z, Moustafa A, Bibi K, Barham A, Abdel-Salam A-SG. Sleep, activity, and cognition: An analysis of academic impact on university students. Acta Kinesiologica. 2024;**18**:24-31

[26] Haddad M, Abbes Z, Abdel-Salam A-SG. The impact of online classes on sleep, physical activity, and cognition functioning among physical education students. Frontiers in Psychology. 2024;**15**:1397588

[27] Erickson KI, Voss MW, Prakash RS, Basak C, Szabo A, Chaddock L, et al. Exercise training increases size of hippocampus and improves memory. Proceedings of the National Academy of Sciences. 2011;**108**:3017-3022

[28] Chtourou H, Souissi N. The effect of training at a specific time of day: A review. The Journal of Strength & Conditioning Research. 2012;**26**:1984-2005

[29] Kross E, Ayduk O. Facilitating adaptive emotional analysis: Distinguishing distanced-analysis of depressive experiences from immersed-analysis and distraction. Personality and Social Psychology Bulletin. 2008;**34**:924-938

[30] Martell CR, Addis ME, Jacobson NS. Depression in Context: Strategies for Guided Action. New York, NY: WW Norton & Co; 2001

[31] Satariano WA, McAuley E. Promoting physical activity among older adults: From ecology to the individual. American Journal of Preventive Medicine. 2003;**25**:184-192

[32] Penedo FJ, Dahn JR. Exercise and well-being: A review of mental and physical health benefits associated with physical activity. Current Opinion in Psychiatry. 2005;**18**:189-193

[33] Tcymbal A, Demetriou Y, Kelso A, Wolbring L, Wunsch K, Wäsche H, et al. Effects of the built environment on physical activity: A systematic review of longitudinal studies taking sex/gender into account. Environmental Health and Preventive Medicine. 2020;**25**:1-25

[34] Mammen G, Faulkner G. Physical activity and the prevention of depression: A systematic review of prospective studies. American Journal of Preventive Medicine. 2013;**45**:649-657

[35] Lee I-M, Shiroma EJ, Lobelo F, Puska P, Blair SN, Katzmarzyk PT. Effect of physical inactivity on major non-communicable diseases worldwide: An analysis of burden of disease and life expectancy. The Lancet. 2012;**380**:219-229

[36] Marquez DX, Aguiñaga S, Vásquez PM, Conroy DE, Erickson KI, Hillman C, et al. A systematic review of physical activity and quality of life and well-being. Translational Behavioral Medicine. 2020;**10**:1098-1109

[37] Wegner M, Helmich I, Machado S, Nardi E, Arias-Carrion O, Budde H. Effects of exercise on anxiety and depression disorders: Review of meta-analyses and neurobiological mechanisms. CNS & Neurological Disorders-Drug Targets (Formerly Current Drug Targets-CNS & Neurological Disorders). 2014;**13**:1002-1014

[38] Trost SG, Owen N, Bauman AE, Sallis JF, Brown W. Correlates of adults' participation in physical activity: Review and update. Medicine & Science in Sports & Exercise. 2002;**34**:1996-2001

[39] Bidzan-Bluma I, Lipowska M. Physical activity and cognitive functioning of children: A systematic review. International Journal of Environmental Research and Public Health. 2018;**15**:800